磷脂酰胆碱在LET-607寿命调控通路中的作用和机制研究

何 斌 著

东北大学出版社
·沈阳·

ⓒ 何 斌 2023

图书在版编目（CIP）数据

磷脂酰胆碱在LET-607寿命调控通路中的作用和机制研究/何斌著. -- 沈阳：东北大学出版社，2023.8
　ISBN 978-7-5517-3379-3

Ⅰ.①磷… Ⅱ.①何… Ⅲ.①卵磷脂-作用-动物-细胞学 Ⅳ.①Q952

中国国家版本馆CIP数据核字（2023）第165619号

出 版 者：	东北大学出版社
	地　址：沈阳市和平区文化路三号巷11号
	邮　编：110819
	电　话：024-83680182（市场部）　83680267（社务部）
	传　真：024-83680182（市场部）　83687332（社务部）
	网　址：http://www.neupress.com
	E-mail:neuph@neupress.com
印 刷 者：	沈阳市第二市政建设工程公司印刷厂
发 行 者：	东北大学出版社
幅面尺寸：	170 mm × 240 mm
印　　张：	11.25
字　　数：	250千字
出版时间：	2023年8月第1版
印刷时间：	2023年8月第1次印刷
责任编辑：	汪子珺
责任校对：	曹　明
封面设计：	潘正一

ISBN 978-7-5517-3379-3　　　　　　　　　　　　　定　价：48.00元

作者简介

何斌，男，汉族，副教授，1989年12月出生于贵州省遵义市。2012年本科毕业于海南大学生物科学专业，2021年12月博士毕业于重庆大学生命科学学院遗传学专业。主持贵州省教育厅服务"四新""四化"科技攻关项目一项（黔教技〔2022〕008号）；主持横向项目"天然橡胶技术开发项目合作协议"（GZSQXSWKJYXGS202201）；主持六盘水师范学院高层次人才启动基金项目一项（LPSSYKYJJ202205），项目经费共计305万元整。作为主要参与人，参与了国家自然科学基金面上项目四项（32071163，32070754，81970008，31771337），国家自然科学基金青年科学基金项目一项（31701017）；作为参与人，参与了国家自然科学基金青年科学基金项目一项（31300570）。指导学生参加第八届全国大学生生命科学竞赛（科学探究类）。发表学术论文11篇，其中SCI收录论文6篇，发明专利6项。

前　言

　　生物体具有复杂的应激反应途径来抵御多种环境和应激压力，而应激反应的紊乱会导致细胞稳态受损和机体衰老。真核细胞进化出膜封闭的亚细胞结构赋予了细胞维持独特微环境的能力，促进了各生物过程的有效间隔，并调节适当细胞功能所需的数千种分子的浓度、转运和扩散。这一膜封闭系统还允许细胞从物理上控制其损伤，从而在特定的亚细胞结构内有针对性地诱导分子伴侣和降解机制，将错误折叠的蛋白质和异常分子隔离起来，使细胞的其他部位仍然正常运转。因此，每个亚细胞结构已经进化出独特的应激能力，当外界施加压力时，应激反应通常在细胞内特定的细胞器中激活，从而使细胞能够特异性应对应激或损伤，同时保持细胞其他部位的正常功能。内质网是蛋白质折叠和脂质生物合成的主要场所，是细胞应激反应（如未折叠蛋白质反应）的关键细胞器，这种蛋白质反应可在蛋白毒性应激（也称为内质网应激）后恢复蛋白稳态。内质网锚定转录因子在内质网应激反应中起关键作用。属于环磷腺苷效应元件结合蛋白3（cAMP-response element binding protein 3，CREB3）转录因子家族的CREBH，是一种重要的内质网结合转录因子。响应内质网应激，CREBH活化后转移到高尔基体，在那里它被蛋白酶进一步裂解，裂解的N末端片段转移到细胞核，以激活下游靶基因，这一过程称为调节膜内蛋白稳态控制。CREBH的激活促进炎症因子和铁调节激素（即铁调素）的表达，从而将内质网应激与细胞防御反应联系起来。

　　本书使用模式生物秀丽隐杆线虫研究哺乳动物细胞防御反应和蛋白

稳态控制反应的调控因子 CREBH 的秀丽隐杆线虫直系同源物 LET-607；揭示了 LET-607 与胞质 DAF-16 应激途径之间的新型交流，即抑制 LET-607 会诱导 DAF-16 的适应性激活，而 DAF-16 是秀丽隐杆线虫中细胞质应激反应的主要调节因子，因此抑制 LET-607 能改善其健康和寿命。这揭示了不同应激途径之间交流对动物适应性的重要性，并确定 LET-607/CREBH 和特定的磷脂酰胆碱是 DAF-16 和寿命的调控因子。本书的研究结果描述了连接不同应激反应途径的细胞机制，并突出了其在动物长寿中的重要性。

本书撰写得到了重庆大学庞珊珊教授和唐海清副教授的鼎力相助，在此感谢两位老师的悉心指导。本书还得到了六盘水师范学院、贵州省教育厅服务"四新""四化"科技攻关项目（黔教技〔2022〕008号），六盘水师范学院学科（培育）团队项目（LPSSY2023XKTD08），六盘水师范学院高层次人才启动基金项目（LPSSYKYJJ202205）资金资助，特此感谢！

著 者
2023年3月

目 录

1 绪 论 ·· 1
  1.1 秀丽隐杆线虫 ··· 2
  1.2 内质网应激反应 ·· 4
      1.2.1 IRE-1 ··· 6
      1.2.2 PERK ·· 7
      1.2.3 ATF-6 ·· 8
      1.2.4 秀丽隐杆线虫的内质网应激反应 ···················· 10
  1.3 DAF-16/FOXO ·· 12
      1.3.1 DAF-16 ··· 12
      1.3.2 FOXO ··· 15
  1.4 CREBH/LET-607 ·· 17
      1.4.1 CREBH ··· 17
      1.4.2 LET-607 ·· 18
  1.5 磷脂酰胆碱 ·· 19
  1.6 立题依据及研究意义 ··· 21
      1.6.1 立题依据 ·· 21
      1.6.2 研究意义 ·· 22

2 材料与方法 ·· 23
  2.1 秀丽隐杆线虫品系及其培养条件 ····························· 23
  2.2 实验室用水 ·· 24
  2.3 秀丽隐杆线虫生长培养基的制备 ····························· 24
  2.4 *OP*50-1 菌株和 RNA 干扰菌株培养皿的制备 ············ 25
  2.5 LB 培养基 ··· 25

2.6	M9缓冲液（M9 Buffer）	25
2.7	S缓冲液（S Buffer）	26
2.8	秀丽隐杆线虫基因组提取液（2X Lysis Buffer）	26
2.9	PBS缓冲液（PBS Buffer）	27
2.10	Hajra's 溶液（Hajra's Solution）	27
2.11	10% Triton™ X-100溶液	27
2.12	50×TAE电泳缓冲液	27
2.13	尼罗红（Nile Red）染料	28
2.14	油红O（Oil-Red-O）染料	28
2.15	秀丽隐杆线虫脂肪尼罗红染料染色	28
2.16	秀丽隐杆线虫脂肪油红O染料染色	29
2.17	秀丽隐杆线虫脂肪油红O染料染色定量	29
2.18	大肠杆菌感受态细胞的制备	30
2.19	秀丽隐杆线虫基因组提取	31
2.20	秀丽隐杆线虫冷冻保存与复苏	31
2.21	秀丽隐杆线虫拍照	32
2.22	秀丽隐杆线虫体长测量实验	33
2.23	秀丽隐杆线虫同步化处理	33
2.24	秀丽隐杆线虫RNA提取	34
2.25	第一链cDNA合成	35
2.26	荧光定量PCR（QPCR）	36
2.27	转录组测序	37
2.28	ChIP-seq分析	38
2.29	气相色谱串联质谱分析法（GC-MS/MS）	38
2.30	脂肪酸分析	39
2.31	薄层色谱法（TLC）	41
2.32	秀丽隐杆线虫应激实验	43
	2.32.1　绿脓杆菌感染实验	43
	2.32.2　活性氧应激实验	44
	2.32.3　热应激实验	44
	2.32.4　内质网应激实验	44
2.33	秀丽隐杆线虫寿命实验	44

2.34　秀丽隐杆线虫产卵量分析 ·· 45
2.35　*let-607p*::GFP::*let-607*转基因线虫构建 ································ 45
　　2.35.1　构建*let-607p*::*let-607*载体 ··· 46
　　2.35.2　构建*let-607p*::GFP::*let-607*载体 ································· 49
　　3.35.3　显微注射 ··· 51
　　2.35.4　整合*let-607p*::GFP::*let-607*载体到线虫基因组 ·············· 51
2.36　化合物补充实验 ·· 52
　　2.36.1　神经酰胺 ·· 52
　　2.36.2　磷脂酰胆碱 ·· 52
　　2.36.3　离子霉素 ·· 53
　　2.36.4　1-油酰基-2-乙酰基-sn-甘油 ··· 53
　　2.36.5　氯化胆碱 ·· 53
　　2.36.6　地昔帕明 ·· 53
2.37　数据量化和统计分析 ·· 54
2.38　主要仪器 ··· 54

3　结　果 ··· 56
　3.1　LET-607对秀丽隐杆线虫应激反应的影响 ································ 56
　　3.1.1　LET-607对线虫先天免疫的影响 ····································· 56
　　3.1.2　LET-607对线虫抵抗外界压力的影响 ······························· 60
　　3.1.3　秀丽隐杆线虫LET-607的细胞定位 ·································· 63
　3.2　LET-607通过DAF-16调控线虫抗压能力和寿命 ······················· 64
　　3.2.1　LET-607对基因表达的影响 ··· 64
　　3.2.2　LET-607通过调节DAF-16入核来激活其功能 ···················· 67
　　3.2.3　LET-607通过DAF-16调节抗压力 ··································· 69
　　3.2.4　LET-607激活DAF-16有助于增加线虫寿命 ······················· 69
　　3.2.5　已知的CREBH调控因子不参与LET-607对DAF-16的调控 71
　3.3　LET-607调控线虫磷脂酰胆碱代谢 ··· 73
　　3.3.1　LET-607对线虫脂肪累积的影响 ····································· 73
　　3.3.2　LET-607对线虫脂肪酸的影响 ·· 74
　　3.3.3　酸性鞘磷脂酶ASM-3不介导DAF-16激活 ························ 78
　　3.3.4　鞘磷脂合酶SMS-5介导DAF-16激活 ······························· 80

      3.3.5 磷脂酰胆碱介导DAF-16激活 ·· 81
      3.3.6 LET-607调控磷脂酰胆碱代谢 ·· 86
      3.3.7 PC介导生殖细胞缺失型突变体中的DAF-16激活 ············· 88
   3.4 LET-607通过钙离子信号调控DAF-16 ··· 93
      3.4.1 ITR-1钙离子通道对DAF-16激活的影响 ························· 93
      3.4.2 钙离子流对DAF-16激活的影响 ······································ 97
      3.4.3 PKC-2和SGK-1参与LET-607介导的DAF-16激活 ············ 99

4 讨 论 ·· 104

参考文献 ·· 107

附 录 ··· 139
   F.1 QPCR引物 ·· 139
   F.2 ChIP-seq结果 ··· 141
   F.3 病原菌感染实验统计结果 ··· 146
   F.4 活性氧应激实验统计结果 ··· 148
   F.5 热应激实验统计结果 ·· 156
   F.6 内质网应激实验统计结果 ··· 161
   F.7 寿命实验统计结果 ·· 162

1 绪 论

细胞在其一生中都面临着各种压力，所以生物进化出复杂的应激反应途径以防御多变的环境和压力，这些压力导致蛋白质的错误折叠和聚集[1]。大量实验数据表明，压力适应和衰老之间有着密切的联系，应激反应的紊乱会导致细胞稳态受损和机体的衰老[2-3]。细胞需要在不断变化的外部条件中监测独立作用的亚细胞结构之间的稳态，在细胞水平上发展防御机制来响应和适应环境中各种压力，以保护整个生物并保持其正常的生长和繁殖的能力[4]。

应激反应通常在细胞内特定的部位被激活，使细胞能够专门应对压力或损害，同时保持其他部分的正常功能。真核细胞进化出膜封闭的亚细胞结构赋予了细胞维持独特微环境的能力，促进了各生物过程的有效间隔，并调节了适应细胞功能所需的数千种分子的浓度、转运和扩散。这一膜封闭系统还允许细胞从物理上控制其损伤，从而在特定的细胞器或亚细胞结构内有针对性地诱导分子伴侣和降解机制，将错误折叠的蛋白质和异常分子隔离起来，使细胞的其他部位仍然正常运转。因此，每个亚细胞结构已经进化出独特的应激能力，在外界对其施加压力时，通过调节不同亚细胞结构的特异性表达基因来应对，这种高度有益于自身环境的网络通常包含数百个基因，称为未折叠蛋白反应（unfolded protein response，UPR），通常仅受单个转录因子的激活来调控，在内质网（endoplasmic reticulum，ER）、细胞质（cytoplasm）和线粒体（mitochondria）的应激反应调节中起着重要的作用[5-9]。但值得注意的是，一旦某一细胞器应对压力的能力受损，可能会对细胞其他部分造成有害的影响，从而扰乱整个细胞的稳态[10-11]。因此每个细胞器仍依赖其他细胞器的适当功能来维持生存[11-13]。在这种情况下，各细胞器之间的交流通信对于确保细胞整体稳态是必不可少的。

此前已有关于细胞在受到外界压力的情况下，协调内质网和线粒体应对的研究。例如，在肥胖症的病理生理条件下，营养物质的过度氧化可能诱导线粒体受到压力，导致线粒体未折叠蛋白反应。UPR^{mt}可能会破坏线粒体的完整性和稳态，从而导致内质网应激[14]；在2020年有研究者发现依赖转录因子ATF-4建立的线粒体与胞质核糖体之间存在逆向通信过程，确定了线粒体的翻译平衡机制，并将核糖体确立为亚细胞器之间的交流通信复合体[15]。研究线粒体与细胞质蛋白稳态反应之间交流的结果表明，线粒体应激可以激活细胞质中的几种蛋白质稳态途径，以恢复细胞稳态[16-18]，并且线粒体tRNA修饰缺陷也可能导致整体细胞蛋白稳态激活机制[19]，这表明线粒体和细胞质之间也存在固有的蛋白质稳态网络。同时，研究者还发现了一种通过脂肪代谢引起脂质动态平衡变化，将线粒体蛋白质动态平衡和胞质应激反应之间联系起来的独特机制，并且将这种反应称为线粒体-胞质应激反应（mitochondrial to cytosolic stress response，MCSR）[20]。

作为蛋白质折叠和脂质生物合成的主要场所，内质网是细胞应激反应的关键细胞器。内质网未折叠蛋白反应（endoplasmic reticulum unfolded protein response，UPRER）是其中重要的应激反应途径，可在应激后帮助内质网恢复蛋白稳态[21]。此外，内质网应激通常与各种环境威胁和内源性损害相关，因此细胞已经进化出整体的内质网应激反应，以恢复整体细胞内稳态[10]。这种整体内质网应激反应通常涉及其他细胞器，内质网和其他细胞器之间的交流通信对于压力条件下的细胞维护非常重要。例如，当内质网受到压力时，内质网未折叠蛋白反应也会对线粒体功能产生一定的影响[10,22]，并影响细胞质的自噬反应[23-24]，从而在内质网应激时维持总体细胞蛋白稳态。细胞器之间的异常交流与蛋白质折叠紊乱导致的疾病的出现和严重程度有关，包括神经退行性疾病、心血管疾病和糖尿病[7,10,25-26]。

目前，除对蛋白稳态控制途径的研究以外，对于细胞内不同的应激途径之间是否存在通信，若存在又如何进行通信以实现总体细胞稳态和生物适应性这一系列问题知之甚少。

1.1 秀丽隐杆线虫

1963年，Sydney Brenner提出Caenorhabditis briggsae这一线虫品种将是解决或将在未来十年内解决关注的分子生物学经典问题的理想系统[27-28]。后来，因为在其实验室秀丽隐杆线虫（Caenorhabditis elegans）相较于其他品种的线虫能够更好地分离和培养，他决定将秀丽隐杆线虫作为他的研究重点[29]。如今秀丽隐杆线虫已被全球上千个实验室用作模式生物开展相关研究。

在世界范围内的土壤中都有秀丽隐杆线虫被发现。新孵化的幼虫长约250 μm，成虫长约1 mm，具有快速的生命周期，并且主要以自体受精的雌雄同体形式存在，雄性的自然出现概率低于0.2%。研究过程中通常可以使用解剖显微镜（可以放大100倍）或复合显微镜（可以放大1000倍）进行观察。解剖显微镜用于观察秀丽隐杆线虫在培养皿上运动、进食、发育、交配及产卵等各生命活动。由于秀丽隐杆线虫是半透明的，研究人员通过使用共聚焦显微镜、微分干涉差显微镜等更精细的仪器设备，可以以更高的分辨率进行观察，从而能够以单细胞分辨率解决与细胞发育和功能有关的科学问题，并通过使用荧光蛋白标记蛋白或亚细胞结构荧光染色技术，更为直观地展示实验过程和结果。荧光蛋白可以用于研究发育过程、筛选影响细胞发育和功能的突变体、分离细胞并表征体内蛋白相互作用等[30-32]。例如，Cameleon和gCaMP3这一类基于荧光蛋白的报道分子会响应钙通量而发出荧光，从而使研究人员能够在荧光显微镜下检测到神经元特异性的钙通量[33]。同时，由于线虫体细胞数量不变，研究人员已经能够追踪活体动物在受精到成年之间每个细胞的变化，并产生完整的细胞谱系[34-36]。

在实验室中，秀丽隐杆线虫被培养在接了大肠杆菌的琼脂培养皿上，并且在12~25 ℃温度范围内培养[37]。在20℃培养时秀丽隐杆线虫从卵中孵出后为第一阶幼虫段（larva 1，L1）。随后线虫开始进食，经历四个幼虫阶段（L1~L4，L1阶段持续约为16 h，其他阶段约为12 h），如图1.1所示。每个阶段的尾声线虫处于类似睡眠状态下的一种不活动的阶段，称为lethargus，此时褪去旧的表皮并形成新的表皮（胶原外层）以满足线虫的外形的生长[38]。L4时期蜕皮约12 h后，成年的雌雄同体开始繁衍后代并持续2~3 d，直到用尽所有自身产生的精子。若雌雄同体耗尽自身精子再与雄性交配，还可以产生更多的后代。生殖期过后，雌雄同体开始逐渐衰老，但仍可存活几周直到死亡。当培养皿上大肠杆菌耗尽，线虫就会消耗自身储存的在肠细胞中的脂肪维持生命活动。而当食物耗尽或者培养皿内线虫数量过多时，L2时期线虫会进入一种被称为dauer的状态，此状态下的线虫被表皮完全包围并塞住其口部阻止其生长发育。这一过程的线虫至少可以存活一个月，而在通常情况下，在15 ℃下可以存活长达六个月之久。当这些处于生长停滞状态的线虫被转移到含有大肠杆菌的琼脂培养皿上时，它们又会重新进入生长发育的生命周期。

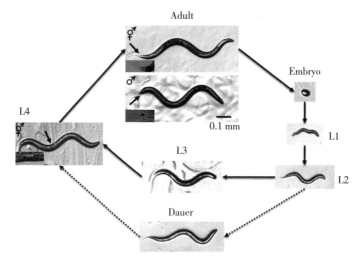

图 1.1 秀丽隐杆线虫生命周期

使用秀丽隐杆线虫作为模式生物进行研究有非常多的优势[39]：第一，秀丽隐杆线虫可以仅由雌雄同体进行自体受精繁衍后代，这样后代基因型明确可控，并且一条雌雄同体从刚孵化到通过自体受精在一周时间内可繁衍约300个后代，短时间内可以得到数量庞大的群体；第二，通过低温冷冻可以长期储存，并且在有需要的时候通过简单操作即可复苏；第三，线虫体积小，这意味着仅需要较小的空间即可繁殖大量实验用线虫；第四，可以在12~25 ℃温度范围内培养，而且温度每增加10 ℃可以提升其2倍生长速度，可以通过温度控制其生长速度，并且可以通过温度敏感型突变体进行简单操作后得到所需表型线虫；第五，通过对妊娠期成虫进行简单操作即可得到大量同步化处理的L1时期幼虫；第六，它们不能在人体体温环境下生长，因此无法对人类造成任何危害，而如猪蛔虫（ascaris suum）之类的线虫会诱发人体过敏反应；第七，线虫是第一个完成完整的基因组序列测序的多细胞生物[40]，60%~80%的人类基因在秀丽隐杆线虫基因组中具有直系同源物[40]，而已知与人类疾病有关的基因中有40%在秀丽隐杆线虫基因组中具有清晰的直系同源物[41]。因此，利用秀丽隐杆线虫可以开展许多与人类健康和疾病有关的研究。

1.2　内质网应激反应

内质网是由囊状、泡状和管状膜结构形成的一个连续的网膜系统，是细胞蛋白质、脂类（如甘油三酯）和糖类合成的场所，同时能对蛋白质进行修饰和折叠，并存储在信号转导中起关键作用的细胞内钙离子。新合成的分泌蛋白和跨膜蛋白易位到内质网中，内质网包含的众多分子伴侣和折叠所需的酶为这些蛋白的折叠提供了最佳环境；同时在内质网分子伴侣的帮助下可以将未折叠或错误折叠的蛋白质逆向转运回细胞质，由Hrd1/Hrd3 E3泛素连接酶复合物泛素化，并通过蛋白酶体介导的内质网相关性降解（ER-associated degradation，ERAD）清除[42]。在某些情况下，末端错误折叠的蛋白质通过ERAD Ⅱ（一种自噬介导的ERAD形式）从内质网中清除[43]。内质网分子伴侣参与的两种机制可确保通过内质网的蛋白质的质量，并仅允许正确折叠的分子沿分泌途径移动[44]。但是，内质网在外在条件下受到损害以及内质网腔内环境发生变化，如腔内钙离子浓度的降低或氧化还原状态的改变，可能会影响蛋白质的折叠和加工。折叠能力降低和错误折叠蛋白在内质网中积累会激活一系列信号通路，这些信号通路统称为内质网应激反应（endoplasmic reticulum stress，ERS）或未折叠蛋白反应（UPR）。几乎所有真核细胞都可以通过激活未折叠的蛋白反应来应对内质网应激并维持内质网的稳态[8]，未折叠蛋白反应可以调节内质网的折叠能力和蛋白质负载，从而促进体内平衡恢复[45-48]。

未折叠蛋白反应最初是在20世纪70年代被发现的，当时对病毒转化的哺乳动物细胞的分析将GRP-78（与主要内质网蛋白伴侣BiP相同）和GRP-94（HSP-90家族的分子伴侣）鉴定为可通过葡萄糖饥饿诱导的蛋白质[49]。而二十世纪八十年代发现内质网中未折叠蛋白的积累可诱导GRP-78和GRP-94的表达[50]。这些发现暗示未折叠蛋白反应是一种稳态反应，通过抑制积累的未折叠蛋白的蛋白毒性来维持内质网中的蛋白折叠环境。在20世纪90年代使用酿酒酵母作为模型来理解未折叠蛋白反应的分子机制，在内质网中存在的未折叠蛋白反应感受器IRE-1（inositol requiring protein 1）以及未折叠蛋白反应特异性转录因子HAC-1的分子克隆方面取得了重大进展[51-54]。随后发现了ire-1介导的hac-1 mRNA剪接，该剪接将内质网与细胞核之间联系起来[52,55]。

内质网应激信号最初是在酿酒酵母中被发现的，其中线性途径仅由应激传感器IRE-1和下游转录因子HAC-1（与哺乳动物ATF/CREB-1同源）调控[8]。内质网未折叠蛋白感受器的数量随着进化而增加，从而使更高的生物体能够以更复杂的方式应对内质网应激[56]。内质网未折叠蛋白反应已发展为一个以多种细胞反应为目标的复杂信号传导网络（见图1.2），它由至少三种主要应激传感器的激活介导：IRE-1（inositol requiring protein 1，α和β亚型），PERK（protein kinase RNA-like ER kinase）和ATF-6（activating transcription factor 6， α和β亚型）[21]。

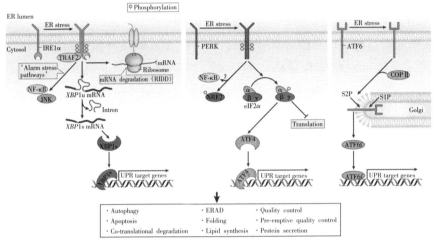

图 1.2　内质网未折叠蛋白反应[21]

1.2.1　IRE-1

　　IRE-1是内质网未折叠蛋白反应中一种高度保守的跨膜蛋白压力感受器，同时也是酿酒酵母唯一的内质网未折叠蛋白反应信号通路。在线虫、果蝇和哺乳动物细胞中IRE-1介导的转录因子从HAC-1转换为XBP-1[57-59]。在哺乳动物细胞中有两种同工型[60]：IRE-1α普遍存在，但IRE-1β仅大量表达在胃肠道上皮细胞中[61-62]。IRE-1的内质网腔参与感知内质网应激[63]，而胞质内核糖核酸酶和激酶结构域则参与将内质网应激信号传递至下游效应子途径。IRE-1途径的激活导致转录因子 xbp-1 受到IRE-1核糖核酸内切酶活性的调控[64]，其剪接 xbp-1 内含子，产生可翻译为有效转录因子的剪接mRNA，称为活化的 xbp-1s，随后被翻译并调节内质网蛋白稳态所需的一系列转录靶标[65-66]。在衰老过程中，内质网伴侣分子被下调，内质网未折叠蛋白反应的激活可能会随着年龄的增长而消失[67]。IRE-1/XBP-1分子的减少影响到秀丽隐杆线虫胰岛素/胰岛素样生长因子1信号途径（insulin/IGF-1 signaling, IIS）的减少，从而对其寿命造成影响[68]。研究者也发现XBP-1s在神经元中的表达启动了细胞非自主反应，从小的透明囊泡（small clear vesicles, SCV）释放神经递质激活了非神经元组织中内质网未折叠蛋白反应，从而防止内质网应激并延长了整个器官的寿命[69]。

　　IRE-1还裂解内质网相关的RNA，导致其衰变，这一反应称为受调控的IRE-1依赖性衰变（regulated IRE-1-dependent decay, RIDD）。研究者通过体外实验发现，IRE-1具有两种不同的作用模式：XBP-1被IRE-1低聚物中协同作用的IRE-1亚单位切割；而IRE-1的单个亚单位在没有协同作用的情况下执行RIDD。此外，使用ire-1 RNase抑制剂STF-083010对XBP-1剪接的选择性抑制表明XBP-1促进细胞存活，而RIDD导致细胞死亡[70]。

IRE-1信号传导受多种不同蛋白质的调控。例如，在成纤维细胞中IRE-1依赖的内质网应激激活由PTP1B（protein tyrosine phosphatase 1B）介导[71]。与胞质小分子热激蛋白的相互作用也可以调节IRE-1的活性，如HSP-72促进IRE-1的核糖核酸内切酶结构域的激活[72]；而与HSP-90的相互作用抑制IRE-1的蛋白酶体降解[73]。多个实验室使用基因芯片或染色质免疫沉淀分析进行的全基因组搜索一致表明，哺乳动物XBP-1不仅参与内质网中蛋白质折叠和降解相关基因的表达，还参与分泌途径相关基因以及其他类似酿酒酵母的基因，表明XBP-1是多功能转录因子[66,74-76]。

1.2.2 PERK

与仅通过诱导转录来应对内质网应激的酵母细胞相反，哺乳动物细胞通常可以通过激活PERK来减弱翻译，从而减轻内质网的负担[77]。PERK最初被鉴定为内质网应激敏感的eIF-2α[78]，PERK是一种I型跨膜蛋白，其N端与IRE-1同源（约20%），C端插入细胞质与蛋白激酶R（double-stranded RNA-dependent protein kinase，PKR）同源（约40%）[79]。与PKR相似，PERK的激酶活性受到寡聚化的刺激[80]，并且在内质网应激后使激酶结构域外的丝氨酸和苏氨酸残基进行自身磷酸化[81]，从而抑制了mRNA转化为蛋白质来调控蛋白质合成以及将新合成的蛋白质折叠成正确的三维结构。这些特性使PERK通过减弱响应ER应激的蛋白质翻译来参与信号转导通路[82]。

在正常情况下，NRF-2（nuclear factor/erythroid-derived 2-like 2）与KEAP-1（kelch-like ECH associating protein 1）结合并定位在细胞质内[83]。然而，在内质网应激之后，PERK响应于内质网应激而磷酸化并激活NRF-2，从而促进其与KEAP-1的解离，并使NRF-2易位至细胞核[83-84]。NRF-2的核定位导致包含抗氧化剂反应元件的基因的NRF-2依赖性转录增加[84]。因此，内质网应激后的NRF-2活性促进细胞存活和氧化还原稳态[85]。

响应PERK激活致使另一种转录因子ATF-4也被激活[86]。哺乳动物的eIF-2激酶PERK被上游应激信号激活时，会抑制大多数mRNA的翻译，但会选择性地增加激活CREB/ATF家族（cyclic AMP response element binding protein/activating transcription factor) ATF-4 的翻译，从而诱导下游基因表达[86-87]。因此，ATF-4 被许多导致 eIF-2α 磷酸化的应激信号通路激活，如氨基酸饥饿和氧化损伤。ATF-4 激活一组常见的应激反应基因，并通过调节细胞代谢活性来促进对内质网应激的适应[88]，响应 ATF-4 激活而表达的成分之一是 bZIP（basic and leucine zipper）转录因子 CHOP（C/EBP homologous protein）[87,89]。包括氨基酸饥饿[90]、遗传毒性和诱变剂[91]和内质网应激[92]在内的许多细胞应激反应都会诱导 CHOP 表达。CHOP 的表达改变了 bZIP 转录因子复合物的结构，以抑制 C/EBP（CCAAT/enhancer-binding proteins）的转录激活[93]，并改善 ATF-3 的转录抑制作用[94]。在许多情况下，CHOP 的诱导被认为是促凋亡的。因为 CHOP 会促进一些凋亡基因的表达，包括 DR-5（cell surface death receptor 5）在内的几种凋亡基因的表达[95]，以及仅 BH3 蛋白 BIM 和 PUMA[96-97]，它们是细胞固有死亡途径中 BCL2 蛋白家族的后一个组成部分。另外，CHOP 诱导生存蛋白 BCL2 的表达降低[98]。CHOP 的表达还可以通过增加 ERO1α 的表达来促进内质网腔的过氧化[98-100]。内质网应激后内质网腔的过氧化促进了肌醇-1,4,5-三磷酸（IP3）诱导的内质网 Ca^{2+} 释放，这归因于 IP3R1（IP3 receptor type 1）的激活[101]。IP3R1 氧化后从与 ERp44 的抑制性相互作用中释放出来，促进了 IP3R1 活化的增强和从内质网腔 Ca^{2+} 的释放，从而促进了细胞死亡[102-103]。因此，CHOP 可以间接地促进内质网应激诱导的内质网中 Ca^{2+} 的释放和细胞死亡。可是在包括小鼠胚胎成纤维细胞（mouse embryonic fibroblast，MEF）的许多类型细胞中，在没有内质网应激的情况下，CHOP 的过表达不足以促进细胞死亡，揭示了需要额外的信号传导，如用 ATF-4 来激活内质网应激诱导细胞死亡[104]。内质网应激后，eIF-2α 磷酸化减弱了整体蛋白的翻译水平。但是，CHOP 和 ATF-4 均可促进 GADD-34（growth arrest and DNA damage 34）的表达，它是蛋白磷酸酶 1 的激活剂，可通过促进 eIF-2α 的去磷酸化来重新激活整体 mRNA 翻译[99,104-106]。

1.2.3 ATF-6

IRE-1 和 PERK 是功能同源的内质网应激传感器，而 ATF-6 则是特异的组成型表达的内质网应激传感器[60]，响应内质网应激而诱导调节性膜内蛋白稳态控制作用转化为活性转录因子[107]，专门用于调节内质网质量控制蛋白体系[108]。酵母基因组中没有 ATF-6，秀丽隐杆线虫和果蝇基因组包含一个 ATF-6 基因，而哺乳动物基因组包含两个密切相关的基因 ATF-6α 和 ATF-6β：ATF-6α 完全负责主要内质网分子伴侣的转录诱导，并且ATF-6α与XBP-1异源二聚体诱导了主要的ERAD成分[109]；与ATF-6α相比，ATF-6β是效率低下的转录激活因子，但它是一种更稳定的蛋白质（活性ATF-6α的半衰期约为2小时，而活性ATF-6β的半衰期为5小时）。另外，在内质网应激后，ATF-6β可以抑制BiP的ATF-6α依赖性转录，这可能是由启动子的竞争所致[110]。

由哺乳动物中密切相关的ATF-6α和ATF-6β组成的ATF-6在内质网中组成性合成的Ⅱ型跨膜蛋白，称为pATF-6α/β（P），其C端位于内质网腔内，N端DNA结合结构域面向细胞质[111-112]。在内质网应激时，pATF-6α/β（P）从内质网易位到高尔基体，并通过site-1 proteases（S1P）和site-2 proteases（S2P）两个蛋白酶的顺序作用而被切割[113-115]。切割后的片段被称为活性形式的pATF-6α/β（N），从膜上释放出来后进入细胞核以激活其靶基因的转录[116-117]，它激活至少三个编码伴侣蛋白的基因（GRP-78，GRP-94和calnexin）的转录，这些蛋白恢复了内质网腔中蛋白质的折叠。ATF-6激活的关键步骤是将其从内质网转运到高尔基体，以便被S1P和S2P切割为有活性的片段。ATF-6在内质网和高尔基体之间的转移是由内质网伴侣BiP控制的，该伴侣是HSP-70伴侣家族的成员，在内质网蛋白折叠和质量控制过程中起着关键作用[114,118-119]。正常情况下，BiP结合到ATF-6的管腔结构域并阻断其固有的高尔基体定位信号，从而将ATF-6隔离在内质网中。当错误折叠的蛋白质在内质网中积累后，ATF-6从与BiP复合物中释放出来，转运至高尔基体。当内质网应激诱导的BiP释放被突变的BiP阻断时，ATF-6的易位和蛋白酶切过程将完全消失。从ATF-6去除BiP结合位点会导致其持续性激活[114,120]。

在哺乳动物细胞中，几乎所有的内质网应激反应基因在启动子区域内均含有一个共有序列（CCAAT-N9-CCACG），该序列充当一些未折叠蛋白反应诱导的转录因子的结合位点[121-122]。N末端结构域是370~380个氨基酸的bZIP家族（basic-leucine zipper）的转录因子，atf-6与xbp-1一起被鉴定为bZIP转录因子，可以结合该共有序列[122]，从而促进包括BiP和CHOP在内的基因的表达[112,123]。有研究结果表明，ATF-6的剪接、核易位或转录激活不需要IRE-1参与，但是IRE-1依赖的未折叠蛋白反应转录诱导需要ATF-6的参与。IRE-1的核糖核酸内切酶活性是剪接xbp-1 mRNA产生新的C末端所必需的，从而将其转换为有效的未折叠蛋白反应转录激活因子。研究者认为ATF-6增加xbp-1 mRNA的数量，而IRE-1剪接XBP-1为有活性的XBP-1s[123]。为了响应内质网应激产生高活性转录因子，xbp-1 mRNA由ATF-6诱导并由IRE-1剪接[59]。

一般情况下，未折叠蛋白反应信号传导可促进生存和对内质网应激的适应，但在蛋白毒性刺激过多的情况下，也可促进细胞死亡。内质网稳态对于正常的细胞功能至关重要，内质网稳态破坏会导致心脏病、神经退行性疾病、糖尿病等[124-126]。此外，许多内质网未折叠蛋白反应相关基因的突变会导致病理变化：XBP-1和SEL1杂合动物会产生胰岛素抵抗[127-128]；小鼠肠道中IRE-1β和XBP-1的缺失会导致肠道炎症和对结肠炎的敏感性[62,129]；XBP-1基因位点的几个单核苷酸多态性是克罗恩病和溃疡性结肠炎的主要危险因素[129]。

1.2.4 秀丽隐杆线虫的内质网应激反应

尽管在哺乳动物中已经对三种未折叠蛋白反应信号转导途径进行了大量的研究，但对于它们如何协调不同靶基因的下游转录、激活以介导在内质网中的蛋白质折叠受到损害时指导适应或凋亡的反应，并且如何参与到其他细胞活动中知之甚少。虽然在酵母中首次发现了IRE-1，但是酵母中不存在PERK和ATF-6信号通路，这限制了通过研究酵母中的未折叠蛋白反应来探讨高级真核生物中该过程的适用性。同时，由于哺乳动物中存在多个IRE-1和ATF-6同源物以及纯合突变的致死性，利用小鼠来进行分析变得复杂；在小鼠胚胎成纤维细胞中的研究无法阐明这些途径的生理和发育功能，因为不同的细胞类型对未折叠蛋白反应子途径有不同的要求[130]。因此，秀丽隐杆线虫成为研究哺乳动物未折叠蛋白反应最好的模型。秀丽隐杆线虫中同时存在三种蛋白可感知内质网应激并激活未折叠蛋白反应：PERK激酶同系物PEK-1、IRE-1和转录因子ATF-6，以及最新发现的LET-607/CREBH[65]。

针对秀丽隐杆线虫三种感受器的类似研究证实了 *ire-1/xbp-1* 和 *pek-1* 突变体表现出最强的表型，并且对衣霉素（tunicamycin, Tm）高度敏感。而与 *ire-1* 和 *pek-1* 不同，*atf-6* 突变体表现出对衣霉素的反应与野生型线虫无差异[65,131-132]。这些结果表明，ATF-6可能在维持内质网和细胞稳态而不是蛋白质折叠方面发挥作用[133]。2005年，研究者基于序列同源性将F45E6.2确定为 *atf-6α* 的唯一线虫同源物，并将其命名为 *atf-6*，秀丽隐杆线虫ATF-6与人ATF-6α具有22%的同源性，并且与人ATF-6β（CREBL1）有约16%的同源性[65]。线虫ATF-6在N端包含一个富含丝氨酸的区域、一个bZIP域和一个由22个可能形成跨膜域的残基组成的疏水片段。C末端包含两个与哺乳动物对应高度同源的区域，这可能是BiP关联和易位到高尔基体所必需的[114]。这些相似性表明线虫ATF-6是一种II型内质网跨膜蛋白，其功能可能类似哺乳动物的ATF-6α。在秀丽隐杆线虫中，*ire-1* 或 *xbp-1* 的缺失会导致胚胎致死，而 *atf-6* 或 *pek-1* 的缺失均在幼虫 L2 阶段产生发育停滞。因此，在秀丽隐杆线虫中，*atf-6* 与 *pek-1* 协同作用，以补充 *ire-1* 或 *xbp-1* 的发育要求。并且秀丽隐杆线虫 *let-607*（CREBH）是由 *ire-1/xbp-1* 和 *atf-6* 调控的新基因，其在人肝癌细胞株 HepG2 中被二硫苏糖醇（dithiothreitol，DTT）诱导的内质网应激反应中瞬时诱导模式与 *xbp-1s* 相似，证实了 *let-607/crebh* 是秀丽隐杆线虫和哺乳动物中的未折叠蛋白反应基因，并且 CREBH 是肝脏特异性转录因子在后生动物进化过程中保守的新型未折叠蛋白反应基因[65]。

减少 IIS 可以增强对压力的抵抗力[134]。*daf-2* 缺失突变的线虫可以抵抗内质网应激，但其 IRE-1 和 XBP-1 活性却相对降低，这可能表明在 *daf-2* 缺失突变中存在与 *xbp-1* 激活无关的通路去改善内质网稳态，这一系统减少了在正常生长条件下以及内质网应激条件下内质网的基础负荷。为此，机体可以将 *daf-2* 缺失突变中的 *ire-1/xbp-1* 途径设置为较低的水平以减轻机体的生存压力[68]。并且存在两个促进内质网动态平衡的独立系统，即使在其中一种途径受到损害时机体也能有效应对内质网应激反应，如当 *ire-1* 介导的未折叠蛋白反应途径受阻时，会出现与衰老相关的现象[69]。随后，研究结果表明独立于保守的 *ire-1/xbp-1* 途径，促进 DAF-16 转录因子向细胞核移位改善了内质网稳态和功能[135]，DAF-16 入核上调的 LGG-1 足以改善 *ire-1* 缺失突变体的内质网稳态和功能[136]。与 *daf-2* 缺失突变情况一致，当保守的 *ire-1* 介导的未折叠蛋白反应途径受阻时，这一系统可以替代经典的蛋白质降解途径，而当经典的未折叠蛋白反应途径和 DAF-16 诱导的途径均完好无损时，这些系统可产生增强的内质网应激能力。与寿命调节的情况一样[135]，核 DAF-16 赋予的内质网稳态改善不如 *daf-2* 缺失突变体。这可能是因为核 DAF-16 转基因的表达水平不足，与 IIS 信号转导减少相关的其他分子事件也可能有助于 *daf-2* 缺失突变体的内质网应激反应。尽管 DAF-16 激活可促进内质网稳态，但内质网稳态本身的改变不足以激活该途径。研究者猜测包括热激和氧化应激在内的其他细胞毒性应激对 DAF-16 的激活可能减弱了通过破坏内质网稳态而直接激活 DAF-16 的进化力。DAF-16 的这种独立于内质网的激活系统可能为跨压力兴奋奠定了基础，甚至在内质网稳态失衡之前就提高了动物应对内质网应激的能力[137]。

1.3 DAF-16/FOXO

1.3.1 DAF-16

在秀丽隐杆线虫早期阶段进行筛选以鉴定调节 dauer 形成的基因，在此筛选中鉴定出的突变称为 daf（dauer formation），分为两类：dauer 组成型（dauer constitutive，daf-c）和 dauer 缺陷型（dauer defective，daf-d）。daf-c 突变体即使在充足的生长条件下也能进入 dauer，而 daf-d 突变体即使在生长条件不利的情况下也无法进入 dauer。在此筛选中，*daf-16* 被分离为 dauer 有缺陷的突变体[138]。遗传上位性分析揭示了 dauer 形成的三种途径：包括 *daf-2*、*age-1*（最初标识为 *daf-23*）和 *daf-16*。*daf-2* 编码的胰岛素/胰岛素样生长因子 1 受体与哺乳动物胰岛素和胰岛素样生长因子 1（IGF-1）受体均相同[139]。*age-1* 编码磷脂酰肌醇（PI）-3 激酶的 p110 催化亚基[140]，而 *daf-16* 编码 FOXO 转录因子[141-142]。*daf-2* 和 *age-1* 最初都被鉴定为 daf-c 突变体，并且需要 *daf-16* 和 *daf-18*。*daf-16*；*daf-2* 和 *daf-16*；*age-1* 双突变体的行为类似于 *daf-16*，表示 *daf-2* 和 *age-1* 通过负调控 *daf-16* 和 *daf-18* 来防止非 dauer 诱导条件下的 dauer 停滞，遗传分析表明 DAF-2 和 AGE-1 可能以单一途径抑制 DAF-16 活性[143-148]。随后表型分析发现 *daf-2* 和 *age-1* 突变体还表现出寿命[149-153]、热应激[154-155]、氧化应激[156-157]、缺氧[158]、渗透压[159]、重金属毒性[160]、紫外线辐射[161]和蛋白毒性[162-164]、脂肪储存[139,165-167]、免疫力[168-169]和繁殖力的变化[170-172]，在 *daf-16* 缺失的突变体线虫中这些功能大都被抑制，即大部分功能都是由 DAF-16/FOXO 介导的。

早期克隆 *daf-2*、*daf-16* 和 *age-1* 基因时，发现它们参与胰岛素/胰岛素样生长因子 1 信号途径（IIS）信号传导途径中。胰岛素和胰岛素样生长因子 1 通过结合并激活具有固有酪氨酸激酶活性的细胞表面跨膜受体而发挥其生物学作用[173-174]。活化的胰岛素和胰岛素样生长因子 1 受体使包括支架蛋白的胰岛素受体底物家族在内的多种底物磷酸化[175]。酪氨酸磷酸化的胰岛素受体底物蛋白促进下游级联反应组分的募集和激活，如磷脂酰肌醇 3 激酶（PI3K）/AKT、促分裂原激活的蛋白激酶（Ras/MAPK）和哺乳动物雷帕霉素靶蛋白途径（mammalian target of rapamycin，mTOR）[176]。这些途径与其他胰岛素和胰岛素样生长因子 1 信号途径在后生动物中广泛保守[177]。此外，IIS 在将系统范围的功能（如生长，繁殖和衰老）与营养状况联系起来方面的作用既得到保护，又具有普遍意义。

秀丽隐杆线虫IIS途径的主要成分包括胰岛素样肽（insulin-like peptides，ILP），其中至少一种可以结合并激活人胰岛素受体[178]。DAF-2/IGFR激活导致磷酸肌醇3激酶AGE-1/PI3K的募集和激活。反过来，丝氨酸/苏氨酸激酶PDK-1，AKT-1和AKT-2被激活，导致DAF-16转录因子的磷酸化。DAF-16的磷酸化调节其与14-3-3蛋白PAR-5和FTT-2的相互作用，后者控制DAF-16亚细胞定位。DAF-18/PTEN脂质磷酸酶和丝氨酸/苏氨酸磷酸酶PPTR-1/PP2A分别抵消AGE-1/PI3K和AKT-1信号转导[179]。通过改变DAF-16（一种FOXO转录因子）、HSF-1（热激转录因子）和SKN-1（一种类似Nrf的异生物反应因子）[180]的表达来抑制IIS信号从而改变寿命，这些转录因子反过来会上调或下调各种基因，这些基因会累积并发挥作用，从而对寿命产生重大影响（见图1.3）。利用遗传易处理系统秀丽隐杆线虫研究IIS的功能非常有价值，对IIS的功能、调节和输出产生了许多重要见解。

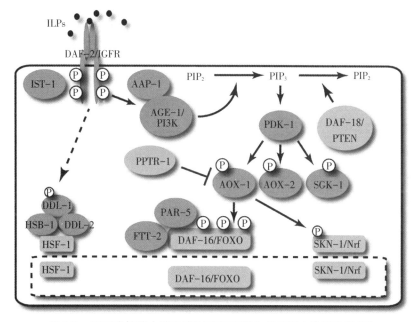

图1.3　秀丽隐杆线虫胰岛素和胰岛素样生长因子1信号途径[179]

据报道，AKT-1在四个不同的位点上磷酸化DAF-16，其中三个在哺乳动物FOXO中是保守的[135,181-182]。在哺乳动物中[183]和秀丽隐杆线虫[184]中血清-糖皮质激素激酶（serum-glucocorticoid kinase，SGK-1）被认为在AKT-1/2的水平上起作用。已在线虫中分离出 akt-1、akt-2 和 sgk-1 三个基因，并且表明它们形成的蛋白质复合物可通过 PDK-1 转导 PI3-激酶信号，从而通过直接磷酸化控制 DAF-16 的定位和激活。有趣的是，早期研究表明 AKT-1/AKT-2/SGK-1 的磷酸化导致 DAF-16 失活[184]；而 2013 年研究者又发现由低温诱导情况下 TRPA-1 介导的 SGK-1 响应钙离子浓度变化而激活 DAF-16，从而使线虫寿命延长[185]。

有丝分裂原激活蛋白激酶（mitogen-activated protein kinase，MAPK）超家族的一个亚家族 Jun 激酶（Jun kinase，JNK），与哺乳动物包括发育、凋亡和细胞存活在内的多种功能有关[186]。JNK 过表达后在线虫[187]和果蝇[188]中直接磷酸化 DAF-16/dFOXO，从而延长了寿命并增加了抗逆性。此外，在秀丽隐杆线虫和哺乳动物细胞培养中，JNK 与 DAF-16 发生物理相互作用并使其磷酸化的位点与 AKT 磷酸化位点不同，并且在秀丽隐杆线虫中与 AKT 作用相比，JNK 增强了 DAF-16 的核易位[187-189]。

AMP 活化的蛋白激酶（AMP-activated protein kinase，AMPK）是具有 α 催化亚基和 β、γ 两个调节亚基的异三聚体激酶复合物，被证明可以在至少六个不同位点直接磷酸化 DAF-16，从而激活 DAF-16[190]。雷帕霉素（rapamycin，TOR）是一种高度保守的激酶，它整合了许多营养信号，并且也与 DAF-16 通路部分重叠[191]，而 AMPK 对 TOR 信号转导的靶标具有抑制作用，因此 AMPK 对 DAF-16 表现出复杂的相互作用。在秀丽隐杆线虫中，TORC2 复合物激活 SGK-1，AMPK 的抑制作用可能会导致 DAF-16 信号的激活[192-193]。此外，最近的一项研究表明，抑制 TORC1 会导致 DAF-16d/f 靶基因表达增加，进而导致转录增加[194]。因此，整个 AMPK 都能直接或间接激活 DAF-16 信号。

此外，TRIM-nhl 基因家族（tripartite motif，TRIM）成员 nhl-1 在线虫神经元中表达会激活远端组织（肠中的 DAF-16）以增强其应激抵抗力[195]。因此，NHL-1 代表由涉及不同组织的 DAF-16 辅因子调节。2014 年研究者通过 RNA-seq 和染色质免疫沉淀（chromatin immunoprecipitation，ChIP）分析全基因组，测序显示目标基因的调控区被 DAF-16 和 SWI/SNF 复合体占据。而且，这些靶基因在很大程度上与包括衰老过程和抗逆性在内的功能有关[196]。SWI/SNF 复合物的成分也在转录调节因子的 RNAi 筛选中被鉴定出来，这些转录调节子通过 DAF-16d/f 延长寿命[197]。这些发现增加了 SWI/SNF 复合体对 DAF-16 的新调节水平。

14-3-3 家族是介导哺乳动物 FOXO 和 DAF-16 核易位的主要参与者。在细胞培养中，AKT-1/AKT-2/SGK-1磷酸化的FOXO与14-3-3蛋白的结合力增加。一旦与14-3-3蛋白结合，FOXO就会从DNA中释放并重新定位到细胞质中[198]。结合的14-3-3将掩盖FOXO上的核定位信号，从而防止其重新进入核内[199-200]。因此，14-3-3蛋白与FOXO的相互作用有助于14-3-3蛋白在细胞核与细胞质间穿梭调节。秀丽隐杆线虫具有两个14-3-3蛋白-PAR-5和FTT-2，并且与哺乳动物类似，线虫PAR-5和FTT-2与DAF-16相互作用，调节其在核与细胞质中的分布[201-203]。sir-2最初在酿酒酵母中被鉴定为一种对基因沉默有重要作用的基因[204]，在哺乳动物中称为SIRT1，与DAF-16类似，现已成为跨系统发育的整个寿命调节的主要参与者[205-207]。而秀丽隐杆线虫14-3-3蛋白还起到介导SIR-2.1（人类同源SIRT1）和DAF-16之间直接相互作用的作用[201-203]。在秀丽隐杆线虫中，过量表达sir-2.1会使dauer形成增加和寿命延长，并且这是由daf-16介导的[207]。这种相互作用也可能会增加DAF-16的特异性。因此，DAF-16的核/胞质穿梭通过与14-3-3蛋白和SIR-2.1的相互作用介导。

秀丽隐杆线虫具有单一的FOXO同源物。DAF-16已成为包括IIS在内的多个上游途径的重要调控因子，响应来自多个途径的信号，DAF-16将结合数千个靶基因，从而转导上游信号以调节不同的生物学过程。

1.3.2 FOXO

Forkhead最初在黑腹果蝇中被分离并命名（现已归类为，Forkhead box A，FOXA），该基因的突变导致异位的头部结构看起来像叉子[208-209]。该家族的特征是保守的DNA结合结构域（Forkhead box，FOX），在人类中有100多个基因，根据序列相似性从FOXA到FOXR分类。这些蛋白质参与多种生物功能：FOXE3对眼睛发育是必需的[210-212]；而FOXP2在语言习得中起作用[213-215]。FOXO蛋白是Forkhead转录因子家族的一个亚族，是进化上保守的转录因子，"O"类成员具有受insulin/PI3K/Akt信号传导途径调控的特征[216]。

FOXO转录因子参与众多关键细胞过程，这些过程包括调控细胞凋亡、细胞周期进程和氧化应激抗性的基因表达程序。例如，FOXO因子可通过激活FasL（Fas依赖性细胞死亡途径的配体）的转录以及激活促凋亡Bcl-2家族成员Bim来启动细胞凋亡[217-219]。另外，FOXO通过上调细胞周期抑制剂p27kip1诱导G1阻滞或上调GADD-45诱导G2阻滞以促进细胞周期停滞[220]。FOXO因子还通过过氧化氢酶和MnSOD的上调影响抗逆性，这两种酶与活性氧的解毒有关[221-222]。此外，FOXO因子可通过上调GADD-45和DDB-1等基因来促进受损DNA的修复[223-224]。其他FOXO靶基因在葡萄糖代谢、细胞分化、肌肉萎缩甚至能量稳态中起作用[219]。在线虫中发现了DAF-16在生物寿命中的关键作用。在哺乳动物细胞中进行的实验表明，FOXO因子是胰岛素和生长因子信号传导的进化保守介体（见图1.4）。

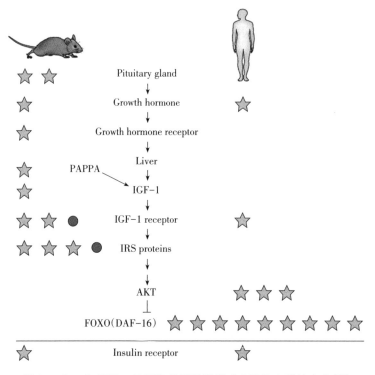

图1.4　Insulin/IGF-1/FOXO信号途径影响小鼠和人类的寿命[134]

FOXO蛋白主要响应胰岛素或生长因子AKT介导的FOXO磷酸化严格调节，三个保守残基的磷酸化导致FOXO因子从细胞核输出到细胞质，从而抑制了FOXO依赖性转录[225]。FOXO蛋白还被包括JNK或Mst1在内的其他蛋白激酶磷酸化，这些激酶在氧化应激条件下将FOXO磷酸化，这种磷酸化导致FOXO从细胞质转移到细胞核，从而起到对抗AKT的作用[226]。除了通过磷酸化进行翻译后修饰外，FOXO蛋白还与共激活因子或共抑制因子复合物结合，并被乙酰化或脱乙酰化。例如，脱乙酰基酶SIRT1通过响应于氧化应激使FOXO脱乙酰基而增加了FOXO与DNA结合能力[227-228]。FOXO蛋白还在氧化应激条件下被泛素化，这增加了其转录活性，最后FOXO蛋白也可以被泛素化并靶向蛋白质降解[229-230]。在特定的环境条件下，FOXO独特的磷酸化，乙酰化和泛素化状态可能在调节FOXO目标基因的子集方面提供特异性。FOXO因子可延长无脊椎动物的寿命。在果蝇中，dFOXO的过度表达足以延长寿命[231]。线虫直系同源物DAF-16可以激活一个基因程序，该程序通过增强对氧化应激、病原体和蛋白质结构破坏的抵抗力来延长寿命。胰岛素受体或PI3K的突变可将寿命延长至三倍，而 *daf*-16 突变后，这种延伸将恢复。缺乏胰岛素受体或胰岛素样生长因子受体1小鼠的寿命比野生型小鼠长30%，这表明FOXO因子可能与哺乳动物长寿有关。此外，在无脊椎动物和哺乳动物之间保守了参与抗逆性的FOXO靶基因，这表明FOXO在机体抗逆性和长寿方面的功能在进化上是保守的。秀丽隐杆线虫提供了在整个生物中研究FOXO的优势，对秀丽隐杆线虫的研究具有在生物水平上评估FOXO的作用的能力，可以将基因的遗传操作（增加或减少水平）直接作为蠕虫的表型结果进行测量。

1.4 CREBH/LET-607

1.4.1 CREBH

环磷腺苷效应元件结合蛋白（cyclic AMP response element-binding protein，CREB）最初被确定为cAMP信号传导途径的靶标。但是对即早期基因（immediate-early gene，IEG）激活的研究表明，CREB是被多种刺激激活的其他信号传导途径的靶标[232]，是最能表征刺激物诱导的转录因子之一，响应包括多肽类激素、生长因子和神经元活性在内的各种刺激，从而被包括蛋白激酶A（protein kinase A，PKA）、促分裂原活化蛋白激酶（mitogen-activated protein kinases，MAPKs）和Ca^{2+}/钙调蛋白依赖性蛋白激酶（Ca^{2+}/calmodulin-dependent protein kinases，CaMKs）在内的蛋白激酶激活。这些激酶都在特定残基丝氨酸133（Ser133）上磷酸化CREB，并且CREB介导的转录也需要Ser133的磷酸化。尽管有这一共同特征，但CREB激活转录的机制仍会因为刺激的不同而异。在某些情况下，CREB的活性和特异性可以通过CREB上其他位点或与CREB相关的蛋白质的磷酸化来进一步调节，从而使CREB在不同的刺激条件下调节不同的基因表达程序[233-234]。

内质网是蛋白质折叠和脂质生物合成的主要场所，也是细胞应激反应的关键细胞器。内质网未折叠蛋白反应在内质网应激时恢复蛋白质稳态[21]，内质网驻留转录因子在内质网应激反应中起关键作用。在哺乳动物中，CREBH是一种内质网结合转录因子，是CREB3家族的成员[235]。Omori等发现CREBH包含一个跨膜结构域，而CREBH的跨膜结构域的缺失增强了瞬时测定中的激活能力。此外，他们还观察到CREBH位于细胞核周围的网状结构中，并在跨膜结构域缺失后观察到该蛋白在细胞核中的定位。基于这些发现，他们提出以下用于调节CREBH转录激活的模型：首先，CREBH锚定在内质网；其次，bZIP结构域和跨膜结构域之间的蛋白酶切反应释放锚定的CREBH蛋白；再次，释放的蛋白质应易位至细胞核并参与转录激活；最后，CREBH在C端包含内质网检索信号，这可能支持CREBH以全长形式定位于内质网的想法。

已有的研究结果表明CREBH调节全身性葡萄糖和脂质代谢[236]。CREBH激活肝糖异生酶磷酸烯醇丙酮酸羧激酶的启动子，而cAMP和蛋白激酶A（PKA）可以进一步刺激这种激活。CREBH转录本在成年肝脏中非常丰富；而在肝癌组织和细胞中，CREBH mRNA的表达异常降低。CREBH的强制表达抑制了培养的肝癌细胞的增殖[237]。CREBH作为肥胖症和糖尿病的治疗靶标，腺病毒和过表达可诱导全身性脂解作用、肝生酮和胰岛素敏感性以及能量消耗的增加，从而导致体重减轻、血浆脂质水平和葡萄糖水平显著降低[238]。并且还激活了抗糖尿病激素（包括成纤维细胞生长因子21和IGF结合蛋白2）的基因表达水平和血浆水平[239]。CREBH与PPARα相互作用，参与脂肪酸的氧化和生酮作用，可能成为非酒精性脂肪肝疾病的治疗靶标[240]。同样，在小肠中高表达的CREBH过表达通过减少Npc1l1表达来预防高胆固醇饮食诱导的高胆固醇血症，预示着可能可以作为高胆固醇血症的治疗目标[241]。

1.4.2　LET-607

LET-607最初在2005年被发现，Shen等鉴定出F57B10.1编码与哺乳动物CREBH同源的bZIP转录因子，随后发现F57B10.1能够调节秀丽隐杆线虫的未折叠蛋白反应。F57B10.1既可以被未折叠蛋白反应激活，也可以促进未折叠蛋白反应[65]。对野生型秀丽隐杆线虫进行衣霉素处理后，线虫的*let-607*上调了约2.73倍。但是，这种上调在*ire-1*和*xbp-1*缺失突变体线虫中被抑制了。并且LET-607的跨膜结构域和ATF-6非常相似，同时*let-607*的瞬时诱导模式与剪接的*xbp-1*相似，这证实了*let-607*是秀丽隐杆线虫和哺乳动物中的未折叠蛋白反应反应基因。2011年Silva通过在秀丽隐杆线虫中建立筛选策略，鉴定出*let-607* RNAi可以防止蛋白质聚集并抑制细胞毒性，并且这一过程不依赖

HSF-1增加折叠细胞的能力，并且该团队首次提及 *let*-607 介导细胞质和内质网应激途径之间的交流通信，但研究者未能进一步阐述相关机理[242]。2013年Guisbert 等使用全基因组RNAi筛选来鉴定对秀丽隐杆线虫热休克反应的正向和负向调节基因，他们发现 *let*-607 可以反向调节 hsf-1 来调控线虫热休克反应[243]。2016年 Weicksel 等通过筛选约37000种化合物质，找到名为 C22（C18H13N6OCl）的小分子化合物，其会使秀丽隐杆线虫卵壳中的一个或多个关键成分加工不当，从而破坏了早期的胚胎发育。进一步研究发现，C22诱导的表型取决于LET-607转录因子及其候选靶基因的上调。LET-607的较高水平导致了参与早期胚胎蛋白质运输的各种靶基因的不适当上调[244]。

目前对LET-607/CREBH的研究结果表明：该蛋白是内质网锚定蛋白，LET-607/CREBH作为最新发现的、独立于其他三条经典内质网应激途径而参与内质网未折叠蛋白反应；在哺乳动物肝脏中参与了葡萄糖和脂质代谢，作为肥胖症、高胆固醇血症、非酒精性脂肪肝疾病和糖尿病的治疗靶标；在秀丽隐杆线虫中可负调控HSF-1从而调节热休克反应，并且介导内质网和细胞质的交流通信而调控细胞解毒等，但是并没有更进一步的机理阐述。

1.5 磷脂酰胆碱

生物为了适应如食物缺乏、环境温度的波动以及生殖状态等不断变化的外界条件，适应性地进化出脂质代谢这一基本机制。脂质对于包括衰老和凋亡在内的多种生物过程至关重要，秀丽隐杆线虫、果蝇、小鼠和人类的脂质谱随年龄增长而产生变化[245-249]。许多长寿命秀丽隐杆线虫和果蝇的突变体中脂质的分布都发生了改变，越来越多的证据表明脂质代谢、寿命调节和生殖状态之间有很强的联系[250]。同时生物产生后代需要能量，脂质可以在多种物种中从体细胞主动转运到性腺细胞和胚胎中[251-253]；而去除性腺可以促进线虫和哺乳动物的脂肪积累并延长其寿命[250]。脂质水平、组成和位置的变化是如何影响寿命的呢？在人和小鼠血浆中发现甘油三酯和脂蛋白质复合物含量随年龄增长而增加[254]。在大鼠衰老过程中，其脑、肝脏和心脏的膜流动性降低[255]；在小鼠肝脏和大脑中膜的磷脂酰乙醇胺（phosphatidylethanolamine，PE）、磷脂酰胆碱（phosphatidylcholine，PC）和鞘磷脂（sphingomyelin，SM）的水平降低，这可能反映了质膜组成的变化[256-257]；而膜组成的改变会改变膜的流动性，当膜的PC和PE比值低、脂肪酸不饱和度高或胆固醇含量低时，膜的流动性就高[255]。

长寿的生物（如裸鼠）表现出特定的磷脂/脂肪酸饱和度特征[258]，这可能有助于维持膜的流动性。此外，诸如饮食限制（dietary restriction，DR）等促进寿命延长的干预措施可防止小鼠的膜流动性因年龄增长而下降[259]。因此，增加的膜流动性可能是促进寿命延长的关键因素。

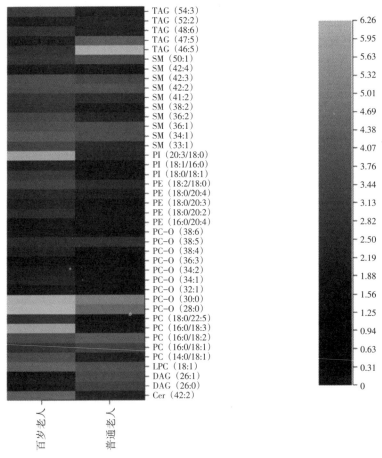

图1.5　普通老人和百岁老人标准化脂质值（μmoL/L）的差异[134]

细胞膜是由两亲性脂质和蛋白质组装而成的高度动态变换的结构，脂质成分赋予质膜重要的性能。膜脂中含量最丰富的是磷脂酰胆碱（PCs），Lin等通过长期施用D-半乳糖建立了大脑加速衰老的小鼠模型，随后对小鼠海马体进行了多平台代谢组学分析，发现磷脂酰乙醇胺（PEs）随着小鼠海马体衰老过程而减少；相反，磷脂酰胆碱（PCs）随着衰老而大量积累[256]。Montoliu等结合核磁共振代谢组学（NMR metabonomics）和鸟枪法脂质组学（shotgun lipidomics）技术对比分析来自意大利的98名百岁老人［平均年龄（100.7±2.1）岁］和196名普通老人［平均年龄（70±6）岁］的血清，发现与普通老人相比，百岁老人两种饱和磷脂酰胆碱水平降低，多种不饱和磷脂酰胆碱水平升高，致使他们饱和/不饱和磷脂酰胆碱的比例增加[260]。这预示着磷脂酰胆碱在生物衰老过程中是一个重要的信号分子，可能起着调节作用（见图1.5）。

1.6 立题依据及研究意义

1.6.1 立题依据

膜脂是细胞内含量最丰富的脂类分子，是生物膜的主要结构组分，主要由磷脂、鞘脂和胆固醇等组成。近年来研究发现，膜脂不仅起着膜结构单元的作用，其组成对细胞信号转导、细胞功能甚至对某些人类疾病也都有非常重要的影响[261-262]。近年来研究者发现，膜脂在衰老调控中可能起着重要的作用。针对人类和模式动物研究显示，衰老伴随着多种膜脂分子含量的变化[263]。比如，膜脂中含量最丰富的磷脂酰胆碱随着年龄增加而增加[256,260]。然而迄今为止，膜脂在衰老调控中的功能和分子机制在很大程度上还不清楚。另外，鉴于膜脂的细胞生物学功能研究尚处起步阶段，解析膜脂的新功能和机制是当前代谢研究领域的前沿和难点。

内质网是膜脂合成的主要场所，是调控细胞脂质合成、维持脂质稳态的重要细胞器。定位于内质网的转录因子胆固醇调节元件结合蛋白（sterol regulatory element-binding protein，SREBP）对胆固醇稳态起调控作用，当胆固醇含量下降时，SREBP转移至高尔基体，蛋白酶S1P和S2P对其进行切割，使其N端在被蛋白酶酶切后释放，入核激活胆固醇合成基因的表达，以维持胞内胆固醇的稳态[264]。这一调控模式反映了内质网进化出精密机制以监督和维持重要脂类的水平，是内质网蛋白调控脂代谢的经典模式。然而，除胆固醇外，其他重要膜脂分子的转录调控机制尚有待进一步阐明。SREBP的激活方式被称为受控膜内蛋白酶解（regulated intramembrane proteolysis，RIP），即跨膜蛋白胞质部分被切割，并作为转录因子入核调控目的基因表达[265]。近年有研究者发现并鉴定出CREB3家族为内质网转录因子家族，同样也以RIP方式激活[266]。CREBH是CREB3家族成员之一，有研究结果表明内质网未折叠蛋白压力以RIP依赖的方式激活CREBH，促进固有免疫基因的表达[267]。本书前期在模式动物秀丽线虫中筛选免疫调控因子时发现CREBH同源蛋白LET-607正调控线虫固有免疫，提示LET-607/CREBH功能具有进化保守性。包括本研究在内的多个研究小组均发现固有免疫与线虫寿命正相关[268-271]，因此利用同样作为衰老研究模式动物的秀丽线虫，本书进一步分析了LET-607对线虫寿命的影响。令人意外的是，尽管LET-607正调控线虫固有免疫，却负调控线虫寿命，*let-607* RNAi能显著增加线虫的寿命。本书推测*let-607* RNAi可能激活了其他重要长寿调控因子。FOXO蛋白是最重要的长寿转录因子之一，其促长寿的作用从秀丽线虫到人类高度保守[272]。

作为转录因子，LET-607可能通过调控目的基因表达来影响衰老进程。本书旨在研究LET-607调控寿命的分子机制，为衰老的代谢调控机制提供新的内容，也将为膜脂的功能提供新的认识。

1.6.2 研究意义

为了探讨LET-607对寿命及各应激反应调控的分子机制，本书以秀丽隐杆线虫为研究对象开展研究。遗传分析发现，*let-607* RNAi激活了线虫FOXO同源蛋白DAF-16，并进一步以DAF-16依赖的方式促进线虫氧化应激抵抗能力的增强及寿命的延长；发现*let-607* RNAi可增加鞘磷脂合酶*sms-5*（sphingomyelin synthase 5）基因的表达，而*sms-5* RNAi则能抑制*let-607* RNAi诱导的DAF-16激活、氧化应激抵抗增强和寿命延长；并且通过生化和遗传分析发现，*let-607* RNAi线虫体内*sms-5*激活导致的不饱和磷脂酰胆碱含量下降是DAF-16激活的主要原因。本书推测LET-607-SMS-5可能代表了一条重要的调控膜脂PC稳态的反馈环路，且该反馈环路与线虫的寿命息息相关。本书揭示了涉及鞘磷脂合酶SMS-5、膜脂质磷脂酰胆碱、内质网驻留钙通道肌醇1,4,5-三磷酸受体ITR-1和钙离子依赖性激酶PKC-2之间交流的潜在机制。总之，这些结果显示了不同应激反应途径之间的明确关系，并突出了其在动物寿命中的重要性，为衰老的代谢调控机制提供新的内容和膜脂的功能提供新的认识。

2 材料与方法

2.1 秀丽隐杆线虫品系及其培养条件

秀丽隐杆线虫遗传中心（Caenorhabditis Genetics Center，CGC）提供了以下线虫虫株：野生型N2，AU78 [*T24B8.5p::gfp::unc-54-3′*UTR]，CF1553 [*sod-3p::gfp*]，TJ375 [*hsp-16.2p::gfp*]，BC10060 [*rCesC12C8.1::gfp + pCeh361*]，CL2166 [*gst-4p::gfp*]，CF1038 [*daf-16*(mu86)]，PS3551 [*hsf-1*(sy441)]，VP303 [*rde-1*(*ne*219); kbIs7(nhx-2p::rde-1)]，WM118 [*rde-1*(*ne*300); neIs9 (*myo-3::ha::rde-1*)]，NR222 [*rde-1*(*ne*219); kzIs9(*lin-*26p::*rde-*1)]，SJ4005 [*hsp-*4::*gfp*]，TJ356 [*daf-*16p::*daf-*16a/b::*gfp*]，CB1370 [*daf-*2(*e*1370)]，DR1572 [*daf-*2(*e*1368)]，CB4037 [*glp-*1(*e*2141)]，JT73 [*itr-*1(*sa*73)]，BX24 [*fat-*1 (*wa*9)]，BX26 [*fat-*2(*wa*17)]，BX30 [*fat-*3(*wa*22)]，BX17 [*fat-*4(*wa*14)]，BX107 [*fat-*5(*tm*420)]，AA18 [*daf-*12(*rh*61*rh*412)]，RG1228 [*daf-*9(*rh*50)]，CF2167 [*tcer-*1(*tm*1452)]，F2052 [*kri-*1(*ok*1251)]，STE68 [*nhr-*49(*nr*2041)]，MGH30 [*sgk-*1(*mg*455)]，MGH266 [*rict-*1(*mg*451)]，BQ1 [*akt-*1(*mg*306)]，VC204 [*akt-*2(*ok*393)]。

陈迪博士提供了DCL569 [*rde-*1(*mkc*36);mkcSi13(sun-1p::rde-1::sun-1 3′ UTR)][273]。

本实验室构建线虫虫株：SSP110 [*glp-*1(e2141);*daf-*16p::*daf-*16a/b::*gfp*]，SSP171 [*mtl-*1p::*mtl-*1::*gfp*]，SSP174 [*let-*607p::*gfp*::*let-*607]，SSP176 [*nhr-*49 (*nr*2041);*mtl-*1::*gfp*]，SSP178 [*akt-*1(*mg*306);*mtl-*1::*gfp*]，SSP179 [*akt-*2 (*ok*393);*mtl-*1::*gfp*]，SSP180 [*sgk-*1(*mg*455);*mtl-*1::*gfp*]，SSP181 [*rict-*1 (*mg*451);*mtl-*1::*gfp*]，SSP182 [*daf-*2(*e*1368);*mtl-*1::*gfp*]，SSP183 [*daf-*2 (*e*1368);*daf-*16p::*daf-*16a/b::*gfp*]，SSP184 [*glp-*1(*e*2141);*mtl-*1::*gfp*]。

秀丽隐杆线虫培养在接种了大肠杆菌*OP*50-1的标准线虫生长培养基上[28]，

根据实验需求的不同置于不同温度的培养箱中。

2.2 实验室用水

本书所述超纯水皆由 Milli-QTM Reference 超纯水系统提供，进水为娃哈哈纯净水，经过第一步纯化，纯水的电阻率即达到 18.2 MΩ·cm（25℃下），且 TOC 值低于 5×10^{-9}，出水为符合 GB/T 6682—2016 规定一级水要求的超纯水。

2.3 秀丽隐杆线虫生长培养基的制备

线虫生长培养基（nematode growth medium，NGM）的制备：每升 NGM 培养基精确称取 3.0 g 氯化钠（Sigma-Vetec，#V900058-500G），2.5 g 蛋白胨（Sigma-Vetec，#V900885-500G），22.0 g 琼脂粉（Sigma-Aldrich，#A1296-5KG）至 1.5 L 三角瓶，加入 970 mL 超纯水后用铝箔纸和橡皮筋封口后，用高压灭菌锅（上海博讯，#YXQ-LS-50A）在 121 ℃下灭菌 50 min。

灭菌结束，待灭菌锅气压降低、温度降至 80 ℃后取出三角瓶，置于提前预热到 55 ℃的水浴锅（科析仪器，#KW-10000DC）中。待培养基温度降低并恒定在 55 ℃后，依次加入提前配置并灭菌处理过的 25 mL 磷酸钾缓冲液（称取 108.3 g 磷酸氢二钾，Sigma-Vetec，#V900050-500G，和 35.6 g 磷酸二氢钾，Sigma-Vetec，#V900041-500G，1L 水溶解后于 121 ℃，灭菌 30 min 后使用），1 mL 硫酸镁溶液（1 mol/L，Sigma-Vetec，#V900066-500G，用 0.22 μm 过滤器过滤后使用），1 mL 氯化钙溶液（1 mol/L，Sigma-Vetec，#V900266-500G，用 0.22 μm 过滤器过滤后使用），1 mL 胆固醇溶液（Sigma-Aldrich，#V900415-25G，5 mg/mL，使用无水乙醇配置，无需灭菌，溶解后即可使用）按照需要再加入对应的抗生素（培养 *OP*50-1 菌用硫酸链霉素，培养 HT115 用羧苄青霉素）以及异丙基硫代-β-D-半乳糖苷（isopropylthio-β-D-galactoside，IPTG，Solarbio，#I8070，贮存浓度 1 mol/L，配置好后用 0.22 μm 过滤器过滤后使用，仅培养 HT115 菌时使用），充分混匀后将三角瓶置于恒温水浴锅中，用灌装蠕动泵（兰格，#WT600-1F）通过灭菌后的硅胶软管将培养基泵入培养皿。为了使各尺寸培养皿内培养基有同一高度，方便在显微镜下观察，90 mm 培养皿泵入 27 mL 培养基，60 mm 培养皿泵入 10 mL 培养基，35 mm 培养皿泵入 4.5 mL 培养基。

2.4 OP50-1菌株和RNA干扰菌株培养皿的制备

将大肠杆菌OP50-1菌株（本实验室保存）接种到含有50 μg/mL硫酸链霉素（Solarbio，#S8290-25g，贮存浓度50 mg/mL，配置好后用0.22 μm过滤器过滤后使用）的LB液体培养基中于37 ℃ 220 r/min下培养过夜，然后将550 μL（根据实际需求增减体积）细菌培养物接种到含有50 μg/mL硫酸链霉素的60 mm NGM培养皿上。在本书实验过程中，如无特殊说明，线虫都喂食OP50-1。

对于RNAi实验，将含有特定序列表达质粒的HT115细菌（购自Ahringer文库）[274]接种到含有100 μg/mL氨苄青霉素（Solarbio，#A8180-25g，贮存浓度100 mg/mL，配置好后用0.22 μm过滤器过滤后使用）的LB液体培养基中于37 ℃ 220 r/min下培养过夜，然后根据需求接种不同体积的菌液到含有50 μg/mL羧苄青霉素（Solarbio，#C8251-10g，贮存浓度50 mg/mL，配置好后用0.22 μm过滤器过滤后使用）和5 mmol/L IPTG的不同尺寸的培养皿上。接种后在25℃下诱导24 h。随后将同步化处理的L1时期线虫添加到RNAi培养皿中培养，以达到RNA干扰指定基因的目的。

let-607 RNAi需将let-607 RNAi菌株的细菌培养物用空载体对照（control RNAi）的细菌培养物以1∶5的比例稀释后使用，pmt-2 RNAi需将let-607 RNAi菌株的细菌培养物用空载体对照的细菌培养物以1∶10的比例稀释后使用。

2.5 LB培养基

精确称取5.0 g氯化钠（Greagent，#G81793D），2.5 g酵母提取物（OXOID，#LP0021），5.0 g胰蛋白胨（OXOID，#LP0043），加水至500 mL于121 ℃下高压灭菌15 min，待温度降至室温后置于4 ℃冰箱，后期根据实验添加所需抗生素后使用。如果需要制备LB固体培养基则每升再加入15.0 g琼脂（Sigma-Aldrich，#A1296-5KG）后灭菌，取出后放置在超净工作台中，并打开紫外灭菌灯，待温度降至56 ℃后于超净工作台中打开，加入相应的抗生素后倒入培养皿中，晾凉凝固后使用。

2.6 M9缓冲液（M9 Buffer）

精确称取5.0 g氯化钠（Greagent，#G81793D），3.0 g磷酸氢二钾（Greagent，

#G82821C)，15.13 g十二水合磷酸氢二钠（Greagent，#G10267D），再加入1 mL 1 mol/L硫酸镁溶液（Sigma-Vetec，#V900066-500G）。加水至1000 mL，充分溶解后于121 ℃下高压灭菌30 min，室温保存。

2.7　S缓冲液（S Buffer）

精确称取5.85 g氯化钠（Sigma-Vetec，#V900058-500G），6.0 g磷酸二氢钾（Sigma-Vetec，#V900041-500G），1.2 g磷酸氢二钾（Sigma-Vetec，#V900050-500G），加水至1000 mL，充分溶解后于121 ℃下高压灭菌30 min，室温保存。

2.8　秀丽隐杆线虫基因组提取液（2X Lysis Buffer）

配置2.5 mol/L氯化钾溶液：精确称取18.6375 g氯化钾（Sigma-Vetec，#V900068-500G），用超纯水定容至100 mL。

配置1 mol/L Tris-HCl溶液：精确称取15.7640 g Tris-HCl（Solarbio，#T8230-500g），加入80 mL超纯水后调节pH值至8.3，最后用超纯水定容至100 mL。

配置0.25 mol/L氯化镁溶液：精确称取5.0825 g氯化镁（Sigma-Aldrich，#M2393-500G），用超纯水定容至100 mL。

配置1%明胶溶液：精确称取0.1 g明胶（Solarbio，#G8061-500g），用超纯水定容至100 mL。

所有溶液配置好后需在121 ℃下灭菌20 min后使用，再加上NP-40（Solarbio，#N8032-100ml）和Tween-20（Solarbio，#T8220-100ml），一次性配置50 mL 2X Lysis Buffer，各组分配比如下：

组分	体积
2.5 mol/L氯化钾溶液	2 mL
1 mol/L Tris-HCl溶液	1 mL
0.25 mol/L氯化镁溶液	1 mL
1%明胶溶液	0.5 mL
NP-40	0.225 mL
Tween-20	0.225 mL
灭菌超纯水	补至50 mL

将配置好的2X Lysis Buffer充分混匀后分装于-20℃冻存备用。

2.9 PBS缓冲液（PBS Buffer）

精确称取8.0 g氯化钠（Sigma-Vetec，#V900058-500G），0.27 g磷酸二氢钾（Sigma-Vetec，#V900041-500G），1.42 g磷酸氢二钠（Sigma-Aldrich，#P5655-100G），0.2 g氯化钾（Sigma-Aldrich，#V900068-500G），加超纯水约850 mL。在磁力搅拌器上充分搅拌，待溶解后将pH值调至7.4，最后定容到1 L，于121℃下高压灭菌30 min，室温保存。

2.10 Hajra's 溶液（Hajra's Solution）

精确称取7.455 g氯化钾（Sigma-Vetec，#V900068-500G），用移液枪取1.2 mL 85%磷酸（Aladdin，#P112025-500ml），加超纯水至100 mL，于121℃下高压灭菌30 min，室温保存。

2.11 10% Triton™ X-100溶液

叔辛基苯氧基聚乙烯乙氧基乙醇（Triton™ X-100，Sigma-Aldrich，#X100-100ML）是一种广泛使用且比较温和的非离子表面活性剂，并且低浓度下对线虫没有杀伤作用，故在实验过程中使用工作浓度为0.01%的Triton™ X-100帮助处理线虫，避免其粘附到离心管、移液枪头等器材上造成损失。

精确量取2 mL Triton™ X-100到50 mL离心管中，再加入18 mL灭菌超纯水制成10%贮存液，振荡混匀。室温静置待泡沫消散后分装到1.5 mL离心管中，于-20℃冻存备用。

2.12 50×TAE电泳缓冲液

精确称取242 g三羟甲基氨基甲烷（Tris，Solarbio，#T0806-500g）和37.2 g二水含乙二胺四乙酸二钠（EDTA·2Na·2H$_2$O，Solarbio，#E8030-500g）于1 L烧杯中，加入850 mL超纯水充分搅拌溶解；加入57.1 mL醋酸（成都科隆，科试），充分搅拌加入超纯水定容至1 L后，室温保存。

2.13 尼罗红（Nile Red）染料

尼罗红（Nile Red，Sigma-Aldrich，#74285-100MG）是一种苯氧嗪染料。可用于秀丽隐杆线虫细胞内脂滴的染色从而进行脂质分析。

精确称取 5 mg 尼罗红到 15 mL 离心管中，加入 10 mL 丙酮（成都科隆，科试）溶液溶解，此为 0.5 mg/mL 贮存液，用铝箔纸完全包裹避光后室温保存，保质期三个月。

2.14 油红O（Oil-Red-O）染料

油红 O（Oil-Red-O，Sigma-Aldrich，#O0625-25G）是一种脂肪染色剂，可用于秀丽隐杆线虫细胞内脂滴的染色从而进行脂质分析。

精确称取 0.5 g 油红 O 染料粉末到 100 mL 蓝盖瓶中，随即加入 100 mL 异丙醇（成都科隆，科试），放入磁力转子后置于磁力搅拌器搅拌混匀。待搅拌混匀后吸出磁力转子，室温保存，保质期六个月。

2.15 秀丽隐杆线虫脂肪尼罗红染料染色

由于线虫产卵会消耗大量脂肪，且虫卵也富含的脂肪会对染色造成影响，同时 *let-607* RNAi 会影响线虫体型，而 L4 时期腹部有非常明显的白色半月型标志，所以本书针对 L4 时期幼虫进行染色。预先用超纯水配置 40% 异丙醇溶液（体积比），以及尼罗红染料工作溶液：取 0.5 mg/mL 贮存液 6 μL 加入 1 mL 40% 异丙醇溶液，避光操作，现配现用。

用含 0.01% Triton™ X-100 的 M9 缓冲液将所需线虫从培养皿上洗到 1.5 mL 离心管中，静置使虫体沉淀到管底。用移液枪小心吸取 M9 缓冲液，随后用 M9 缓冲液反复清洗虫体 3 次，最后尽可能吸净液体。

每个样品加入 800 μL 40% 异丙醇溶液重悬虫体。静置约 5 min 后于 3000 r/min 离心 1 min。吸净上清液，再加入 800 μL 尼罗红工作溶液染色，此过程需全程用铝箔纸包裹离心管以避光反应。

约 2.5 h 后（时间不宜太长），于 3000 r/min 离心 1 min。吸净液体后加入 800 μL 含 0.01% Triton™ X-100 的 M9 缓冲液浸泡半小时，半小时后再用含

0.01% Triton™ X-100的M9缓冲液反复清洗虫体3次，最后吸出染色后的线虫拍照。

2.16 秀丽隐杆线虫脂肪油红O染料染色

由于线虫产卵会消耗大量脂肪，且虫卵也富含的脂肪会对染色造成影响，同时 let-607 RNAi 会影响线虫体型，而L4时期腹部有非常明显的白色半月型标志，所以本书针对L4时期幼虫进行染色。预先用超纯水配置60%异丙醇溶液（体积比），以及油红O工作溶液：60%油红O溶液需用0.22 μm过滤器过滤后立即使用。

用含0.01% Triton™ X-100的M9缓冲液将所需线虫从培养皿上洗到1.5 mL离心管中，静置使虫体沉淀到管底。用移液枪小心吸取M9缓冲液，随后用超纯水反复清洗虫体3次，最后尽可能吸净液体。

每个样品加入800 μL 60%异丙醇溶液重悬虫体。静置约5分钟后于3000 r/min离心1 min。吸净上清液，再加入800 μL 60%油红O工作溶液染色。

约8 h后（时间不宜太长），于3000 r/min离心1 min。吸净上清液，用含0.01% Triton™ X-100的M9缓冲液反复清洗虫体3次，最后吸出染色后的线虫拍照。

2.17 秀丽隐杆线虫脂肪油红O染料染色定量

此实验方法参考上海交通大学徐晓颖博士的博士学位论文，结合本实验室实际情况进行优化。

对染料萃取并定量需要4000只L4时期的幼虫。将染色后的线虫加入1 mL含0.01% Triton™ X-100的M9缓冲液，通过涡旋振荡器将虫体菌液悬浮于液体中，再取10 μL液体在显微镜下数出含有的虫体数量，此步骤需得到每次差异不可太大的三次有效数据以保证精度，取平均值。

将每个样品取相同虫体放置新的1.5 mL离心管中，于3000 r/min离心1 min。吸净上清液，再加入700 μL 100%异丙醇，盖严盖子并用石蜡膜封口后于37 ℃恒温2 h萃取虫体内油红O染料。

在萃取结束前15 min用油红O染料配置0，0.78125，1.5625，3.125，6.25，12.5，25，50，100 ng/μL的标准曲线。萃取结束后将离心管于3000 r/min

离心 1 min，与标准品一起各取 200 μL 到酶标板，在 490 nm 波长下测其吸光值，包括标准品在内的每个样品需做 3 个技术重复。

通过标准品的吸光值制作标准曲线，并且通过使用 100% 异丙醇的用量来调整未知样品的吸光值在标准样品 3.125，6.25，12.5，25 ng/μL 吸光值之间。最后通过标准曲线求得的方程式计算出未知样品的油红 O 染料浓度，以此得到样品的脂肪含量。

2.18 大肠杆菌感受态细胞的制备

本书用到两种感受态细胞，分别是 *DH*5α 和 *HT*115。两者制备方法一致。

将两种大肠杆菌分别接种到无抗生素 LB 液体培养基中，于 37 ℃ 220 r/min 培养过夜，次日早上将过夜培养的菌液按照 1∶100 的体积比接种到 200 mL 新鲜的无抗生素 LB 液体培养基中，于 37 ℃ 220 r/min 培养，约 2.5~3.0 h 后取出 1 mL 细菌培养物，用新鲜 LB 作为对照将两者各取 200 μL 到酶标板中，再将酶标板置于全波长酶标仪（Molecular Devices，#SpectraMax 190）中，在 OD600 下测其吸光值，当其 OD 值介于 0.3~0.4 时立即将细菌培养物取出插入冰上快速降温停止其生长。

待温度降低后，将细菌培养物分批次置于提前预冷至 4 ℃ 的低温离心机（Sigma，#3K15）中，于 4 ℃ 3000 r/min 离心 3 min 收集菌体。

弃上清，加入 10 mL 提前插于冰上的 0.1 mol/L $CaCl_2$ 溶液（Sigma-Vectc，#V900266-500G，0.22 μm 过滤器过滤除菌），用移液枪轻柔吹打重悬菌体，于 4 ℃ 3000 r/min 离心 3 min 收集菌体。

弃上清，加入 10 mL 提前插于冰上的 0.1 mol/L $CaCl_2$ 溶液和 10 mL 提前插于冰上的 30% 甘油（Sigma-Aldrich，#G5516-500ML，用超纯水按体积比配置，121 ℃ 灭菌 20 min 后 4 ℃ 保存）溶液，用移液枪轻柔吹打重悬菌体，再分装到提前预冷的 1.5 mL 离心管中，-80 ℃ 冻存。

次日分别在 *DH*5α 中转入 1 ng pPD95_79 质粒，在 *HT*115 中转入 1 ng L4440 质粒，同时做空白对照。随后涂布含有 100 ng/mL 氨苄青霉素的 LB 固体培养基，37 ℃ 培养箱倒置培养 16 h 后检测所制感受态细胞转化效率。

2.19　秀丽隐杆线虫基因组提取

根据实际实验需求提取各基因型秀丽隐杆线虫基因组。

用含0.01% TritonTM X-100的M9缓冲液将所需线虫从培养皿上洗到1.5 mL离心管中，静置，使虫体沉淀到管底。用移液枪小心吸取M9缓冲液，随后用超纯水反复清洗虫体3次，尽可能清洗干净随虫体一起洗下的菌液。

在200 μL离心管中加入19.6 μL 2X Lysis Buffer和0.4 μL蛋白酶K（Merck，#1245680100，10 mg/mL），再吸入20 μL含有10条左右的线虫到裂解液体系中，经过PCR仪（analytic-jena，#EasyCycler 96）反应后可得到40 μL基因组。所得基因组需-20℃保存，保质期一个月。反应温度和时间如下：

反应温度	反应时间
65 ℃	90 min
95 ℃	15 min
16 ℃	∞

2.20　秀丽隐杆线虫冷冻保存与复苏

根据实际需求及时将线虫冷冻保存和复苏，以免造成某些珍贵品系线虫的丢失。

将约300只L1时期同步化处理后的线虫转移到接种了550 μL OP50-1的60 mm培养皿上，将培养皿置于对应品系线虫所需生长温度。待线虫长到day 2期成虫，此时皿内成虫已进入大量产卵期，将成虫洗下丢弃，此时皿上有非常多的卵和残留非常少的OP50-1。经过约20 h，此时所有的卵都孵化并且L1幼虫吃尽皿上残留OP50-1进入L2时期；再等待约10 h，所有的幼虫饿一段时间后大部分会进入dauer时期，此时即为最佳的冷冻保存时期。每次冻存需同时准备4皿以保证成功率。

用含0.01% Triton X-100的S缓冲液将所需线虫从培养皿上洗到1.5 mL离心管中，3000 r/min离心1 min使虫体沉淀到管底。

离心过程中在准备好的冻存管内加入500 μL 30%甘油（Sigma-Aldrich，#G5516-100ML，用超纯水按体积比配置，121 ℃灭菌20 min后，4 ℃保存备用），待离心结束后用移液枪吸取离心管底部 500 μL 虫体加入冻存管中，与甘油混合均匀置于异丙醇冻存盒内（提前在盒内加入异丙醇，且每两周一换），将冻存盒放入 −80 ℃ 超低温冰箱，在 24 h 后将冻存管从异丙醇冻存盒中取出放入普通冻存盒中长期保存。

约 1 个月后取出四个冻存管中任意一管，于室温放置，待溶液刚好完全溶解（切不可静置太长时间，时间越长复苏率越低）时用移液枪将底部 200 μL 虫体吸出，转移到一个只接种了 200 μL *OP*50-1 的 60 mm 培养皿上，注意带有虫体的液体不要直接转移到菌斑上，应该围绕菌斑接种。幼虫如果复苏则会爬到菌斑上进食，从而造成明显爬痕，这样在第二天通过观察菌斑就能够很轻易判断是否有复苏的幼虫以及查看复苏率。

2.21 秀丽隐杆线虫拍照

用超纯水配置 100 mmol/L 左旋咪唑盐酸盐（Sigma-Aldrich，#31742-250MG）溶液，−20 ℃ 冷冻备用，使用前用超纯水稀释至 2 mmol/L 使用，现配现用。

先用超纯水配置 2% 琼脂糖，用微波炉加热至琼脂糖完全溶解。取载玻片六片，左右各两片叠加，中间只放一片，用移液枪取 60 μL 2% 琼脂糖滴到中间单片载玻片上，另取一片载玻片迅速盖到液滴上。此时两片载玻片会将 2% 琼脂糖液滴挤压为一个有一定厚度的圆形琼脂糖垫片，可在后续实验较长时间为活体线虫提供湿润的环境，保障线虫的水分流失不会过快；且后续使用挑虫器（picker）在琼脂糖垫片上将线虫排列整齐的时候不会造成垫片厚薄不均导致对焦模糊。

含 0.01% Triton X-100 的 M9 缓冲液将所需线虫从培养皿上洗到 15 mL 离心管中，静置让虫体自然沉降后移除上清液，每管中加入 2 mmol/L 左旋咪唑盐酸盐溶液 100 μL，用移液枪吹打混匀，静置 2 min，待线虫麻痹后沉于离心管底部。

用移液枪将离心管底部虫体移至琼脂糖垫片上，然后用挑虫器将散落的虫体聚集到一起。如果短时间内完成拍照则不需要盖载玻片（大部分情况下使用载玻片会造成各虫体间大量气泡产生而对拍照造成影响）。

本书使用的荧光显微镜拍摄系统为尼康产品，其中荧光显微镜 ECLIPSE Ci，光源 Intensilight C-HGFI，显微镜相机 DS-Fi1c，显微镜相机控制器 Digital

Sight DS-U3，软件 NIS-Elements D 4.30.00。

2.22 秀丽隐杆线虫体长测量实验

将所需测量体长的线虫用含 0.01% Triton X-100 的 M9 缓冲液将所需线虫从培养皿上洗到 15 mL 离心管中，静置让虫体自然沉降后移除上清液，每管中加入 2 mmol/L 左旋咪唑盐酸盐溶液 100 μL，用移液枪吹打混匀，静置 2 min，待线虫麻痹后沉于离心管底部。用移液枪将离心管底部虫体移至琼脂糖垫片上，然后用挑虫器将散落的虫体聚集到一起后拍照。

本书使用软件 NIS-Elements D 4.30.00 内嵌长度测量工具，进行线虫体长测量，每个样品测量 120 条线虫后进行统计分析。

2.23 秀丽隐杆线虫同步化处理

为方便后续实验和分析，将线虫进行同步化处理得到同一时期的幼虫。

线虫进入 day 1 成虫时期即开始产卵，并且在 day 2 进入大量产卵时期，所以为了尽可能多的得到同步化 L1 时期幼虫，本书都使用 day 2 时期成虫进行同步化处理。

提前配置 2X 裂解液，配置 1 mL 时各组分配比如下：

组分	体积
84 消毒液	0.2 mL
10 N 氢氧化钠	0.115 mL
超纯水	0.685 mL

根据实际需求现配现用。

用含 0.01% Triton X-100 的 M9 缓冲液将所需线虫从培养皿上洗到 15 mL 离心管中，最后将 M9 体积调为 1 mL，则每个样品用所需 2X 裂解液体积为 1 mL，此时裂解液工作浓度为 1X，充分混匀后静置 3~5 min（可用涡旋振荡器震荡加速裂解）。

在显微镜下观察到有超过四分之三的线虫表皮都破损之后，立即加入 M9 缓冲液到 15 mL 刻度线，随即用 3000 r/min 离心 1 min 收集虫卵。小心倒掉上清液，再用新鲜的 M9 缓冲液反复清洗 3 次后，最后加入 8 mL M9 缓冲液，盖紧离心管盖子将离心管置于混匀仪上以20 r/min的速度于室温过夜孵化。

次日先在显微镜下观察，如果线虫大部分孵化即可将离心管用3000 r/min的速度离心1 min收集幼虫，然后根据实际实验需求将已同步化的L1时期幼虫转移到对应的培养皿中。

2.24 秀丽隐杆线虫RNA提取

根据实际需求提取各实验处理后秀丽隐杆线虫RNA。实验过程中所用离心管、移液枪头等物品都需使用无RNA酶（RNase-free）产品。

使用0.01%焦碳酸二乙酯（Diethyl Dicarbonate，DEPC，生工生物，#B600154-0025）水溶液浸泡处理研磨棒，同时准备50 mL DEPC处理的超纯水，加入磁力转子后置于磁力搅拌器搅拌混匀。处理24 h后于121 ℃高温灭菌20 min后晾凉放入4 ℃冰箱预冷备用。

用含0.01% Triton X-100的M9缓冲液将所需线虫（约2000只L4时期幼虫）从培养皿上洗到1.5 mL离心管中，静置使虫体沉淀到管底。用移液枪小心吸取M9缓冲液，随后用DEPC处理水反复清洗虫体3次，尽可能清洗干净随虫体一起洗下的菌液，并且在最后一步尽可能吸净残留DEPC处理水。这一步需要注意的是，若验证RNAi效率，则需考虑构建RNAi菌时所使用该基因的片段；后续设计的QPCR引物应避开这些片段；若因为基因太短无法避开，则应将线虫置于干净的无任何食物的培养皿上20 min，使线虫消耗干净肠道内残留的细菌以保证后续QPCR的准确性。

随后，在离心管中加入50 μL Trizol试剂（生工生物，#B610409-0100）浸没虫体，将离心管置于冰上并使虫体部分完全淹没在冰下方，在研磨器上套入研磨棒后短时间多次数的研磨虫体（避免长时间研磨导致摩擦产生高温造成RNA降解），研磨后加入1 mL Trizol试剂并将离心管置于冰上。

10 min后，在管内加入200 μL三亚甲基氯溴化合物（1-Bromo-3-chloro-propane，1-BCP，Sigma-Aldrich，#B9673-200ML），轻柔颠倒混匀后置于冰上15 min，待其中液体分层，在静置期间提前预冷离心机到4 ℃。

静置时间到后将离心管取出置于冷冻离心机中，在4 ℃下用13500 r/min的速度离心15 min。离心结束后取出离心管，小心吸出约600 μL上清液至新的1.5 mL离心管中，加入等体积异丙醇，轻柔颠倒混匀后置于-20℃过夜。

第二天取出离心管，置于提前预冷的离心机中，在4 ℃下用13500 r/min的速度离心 20 min，注意此步骤离心管放置到离心机中应使管体与管盖连接处朝外，这样离心后方便吸出上清液而不会触碰到 RNA 沉淀，离心等待期间用 DEPC 处理水将无水乙醇配置为 70%乙醇。

待离心结束后可看到离心管底部有白色沉淀，此即 RNA 沉淀，小心并尽量吸净上清液，加入 500 μL 配置好的 70%乙醇，轻柔颠倒混匀立即置于预冷的离心机中，在 4 ℃下用 13500 r/min 的速度离心 10 min，此步骤离心管放置方向与上一步骤一致。

离心结束后小心吸出上清液，注意此步骤需使用 200 μL 移液枪头分多次将上清液吸出，并且一个移液枪头不可多次使用，最后使用 10 μL 移液枪头吸出残留上清液，此操作步骤目的为避免使用过的移液枪头重新伸入离心管带入移液枪头外壁沾到的 70%乙醇到离心管壁，导致乙醇污染 RNA。

吸净上清液后，将离心管置于 4 ℃冰箱中放置 15 min，此步骤使残留肉眼不可见的乙醇挥发。随后根据 RNA 沉淀的量加入适量的 DEPC 处理水溶解 RNA，再用 Nanodrop 2000 检测 RNA 的浓度、OD260/280 以及 OD260/230，确保 OD260/280 > 1.8 和 OD260/230 > 1.5 后，再取 1 μL 样品用 2%琼脂糖凝胶电泳（北京六一，DYY-7C）检测 RNA 的完整性。将检测合格的 RNA 分装后于 −80 ℃ 超低温冰箱保存备用。

2.25　第一链 cDNA 合成

根据实际需求将提取后秀丽隐杆线虫 RNA 反转录得到 cDNA。实验过程中所用离心管、移液枪头等物品都需使用无 RNA 酶产品。

根据 Nanodrop 2000 实际测得的各样品 RNA 浓度，使用总质量 1.5 μg 的 RNA，精确吸出对应体积的 RNA 用无 RNA 酶的 DNase I（Thermo ScientificTM，RNase-free，#EN0521）进行 DNA 消化。根据说明书，在 200 μL 离心管中逐步加入如下组分：

组分	体积
RNA	1.5 μg
10X 反应缓冲液	1 μL
DNase I	1 μL
DEPC 处理水	至 10 μL

各组分加入后用移液枪头吹打,充分混匀后将离心管置于PCR仪中,于37 ℃下孵育反应30 min,随后加入1 μL 50 mmol/L EDTA溶液后再次放入PCR仪中,于65 ℃下孵育反应10 min使DNA酶失活,反应完后立即取出离心管插入冰上使温度降低。

待温度降低后,使用RevertAid First Strand cDNA Synthesis Kit（Thermo ScientificTM, #K1622）试剂盒,根据说明书操作。将离心管继续插入冰上往管内加入1 μL Oligo（dT）18引物,随后将离心管置于PCR仪中,于65 ℃下孵育反应5 min。反应结束后立即取出插到冰上,待温度降低后再依次加入以下组分：

组分	体积
5X 反应缓冲液	4 μL
Ribolock RNase Inhibitor	1 μL
10 mmol dNTP Mix	2 μL
RevertAid M-MuLV RT	1 μL

各组分加入后用移液枪头吹打充分混匀后将离心管置于PCR仪中,于42 ℃下孵育反应60 min,随后于70 ℃下孵育反应5 min,反应完后立即取出插入冰上使温度降低。然后加入280 μL灭菌后超纯水将cDNA稀释15倍后置于-20 ℃保存备用,尽量在一个星期内使用。

2.26 荧光定量PCR（QPCR）

根据实际需求将提取后秀丽隐杆线虫RNA反转录得到cDNA进行QPCR实验。实验使用iTaq™ Universal SYBR® Green Supermix（Bio-Rad, #1725121）,根据说明书可知所需引物浓度为0.3~0.5 μmol/L,后续将QPCR引物、正反引物等比混合后稀释至2 mmoL/L,10 μL体系加入混合后引物2 μL即可。

在0.2 ml 8-Tube PCR Strips（Bio-Rad, #TLS0851）中依次加入以下组分：

组分	体积
2X iTaq™ Universal SYBR® Green Supermix	5 μL
引物	2 μL
cDNA	3 μL

将各组分添加后,用Optical Flat 8-Cap Strips for PCR Tubes（Bio-Rad,

\#TCS0803)盖严,在涡旋振荡器上振荡以充分混匀,短暂离心后将样品置于仪器上运行(Bio-Rad®,CFX96TM,#1855195),运行程序如下:

模式	温度	时间	备注
聚合酶激活和预变性	95 ℃	20 s	
变性	95 ℃	3 s	
退火/延伸和数据读取	60 ℃	15 s	回至变性,循环39次
熔解曲线分析	65~95 ℃	每2 s升高0.5 ℃	

运行结束后使用CFX Maestro™ Software(Bio-Rad®,#12005258)进行分析,得到各样品CT值,最后采用参照基因的ΔCT法来进行计算:

$$\text{Ratio(reference/target)} = 2^{\text{CT (reference)} - \text{CT (target)}}$$

在此前研究中常用的,并且在各处理下线虫中无明显变化的基因 snb-1 为本书所用参照基因。对每种样品中的目的基因的表达量进行归一化后,再计算相对表达量。

本书所涉及的荧光定量PCR引物均自行设计后送生工生物工程(上海)有限公司合成,纯化方式为PAGE。涉及的引物详见附录F.1。

2.27 转录组测序

根据实际需求,对野生型线虫N2进行control RNAi和 let-607 RNAi,提取线虫总RNA,经过实验室检测RNA质量合格后送美吉生物公司用高通量测序技术进行测序分析。实验过程中所用离心管、移液枪头等物品都需使用无RNA酶产品。

美吉生物使用Agilent Bioanalyzer 2100(Agilent)和ND-2000(NanoDrop Technologies)检测RNA样品,在样品参数达到OD260/280=1.8~2.2,OD260/230≥2.0,RIN≥6.5,28S∶18S≥1.0,总质量大于10 μg后才能用来构建测序文库。该文库在Illumina HiseqX 10上以配对末端150 bp的阅读长度进行测序。使用Tophat v.2.0.6[275]将RNA-seq数据与参考基因组WBcel235(参考基因组来源: http://metazoa.ensembl.org/Caenorhabditis_elegans/Info/Index)进行比对。使用Cufflinks工具进行差异基因和转录本表达分析[276-278]。倍数变化大于2(上调或下调)且FDR小于0.1的基因被视为差异调节基因。使用DAVID(Database for Annotation, Visualization, and Integrated Discovery,版本6.8)的数据

库进行基因功能分类和GO术语分析[279]。

RNA测序数据已保存在GEO中，登录号为GSE155935。

2.28　ChIP-seq分析

ChIP-seq分析数据在2016年被研究者提交到NCBI上，登录号为GSE84419的三个样品GSM2233444（LET-607 ChIP in staged YA rep 1），GSM2233445（LET-607 ChIP in staged YA rep 2），GSM2233446（Input DNA control in staged YA）[244]。数据分析由武汉爱基百客生物科技有限公司进行。

使用Trimmomatic（0.38版）过滤低质量的读数[280]。使用Bwa（0.7.15版）将干净的读数映射到秀丽隐杆线虫基因组（WBcel235, http://metazoa.ensembl.org/Caenorhabditis_elegans/Info/Index）[281]。Samtools（1.3.1版）用于删除潜在的PCR重复项[282]。使用MACS2软件（2.1.1.20160309版）通过默认参数（带宽300 bp，模型折叠数5、50，q值0.05）调用峰。如果一个峰的峰值最靠近一个基因的TSS，则该峰将被分配给该基因[283]。使用IDR（不可重复发现率）鉴定高可信度结合事件，使用5%IDR分数阈值获得交流复制阈值。

此次分析共获得1133个peak（附录F.2），利用最新技术和算法进行的分析结果与此前的结果进行对比，如 *emo-1*、*enpl-1*、*sec-61* 等基因此次分析的结果中也是存在的，因此本次数据分析结果真实可靠，可用于后续研究。

2.29　气相色谱串联质谱分析法（GC-MS/MS）

实验前处理方法参考美国著名脂类代谢科学家Jennifer Watts博士的文章[284]，在不影响最终分析的情况下，结合本实验室硬件条件进行适当修改。本书甲酯化脂肪酸标准品为Supelco® 37种组分 FAME 混标（Sigma-Aldrich, #47885-U)，购买的标准品总共有37种甲酯化脂肪酸，在重庆大学分析测试中心郑国灿老师、岛津（成都）办事处包晓明工程师以及岛津（广州）检测技术有限公司康文昱工程师的帮助下建立了甲酯化脂肪酸Q3 Scan全扫描MRM方法。实验过程中所用试剂均为色谱纯，所涉及容器均为玻璃或聚四氟乙烯（poly tetra fluoroethylene，PTFE）材质。

使用仪器为岛津GCMS-TQ8040三重四极杆型气相色谱质谱联用仪；色谱柱为岛津SH-Rxi-5sil (Shimadzu, #221-75954-30, 30 m × 0.25 mm)；99.999%氦

气作为载气,速度为1.4 mL/min;进样口温度280 ℃,不分流;火焰离子化检测器(flame ionization detector, FID);将气相色谱仪的初始温度设置为40 ℃保持2 min,然后以6 ℃每分钟的速度升至320 ℃;使用EI离子源轰击样品,离子源温度200 ℃,界面温度280 ℃。将此方法命名为FA_ME_DB5MS_EI_V3_Scan。

将37种组分FAME混标用色正己烷等梯度稀释为10×10^{-6},5×10^{-6},1×10^{-6},0.1×10^{-6},0.01×10^{-6}五个浓度。首先用Q3 Scan全扫描10×10^{-6}标准样品,加载FA_ME_DB5MS_EI_V3_Scan方法,待仪器检测完成,再根据标准品说明书提供的各组分保留时间,在所得结果范围内找到对应的峰,通过检索本地数据库精准识别出目标峰,并定位其精确保留时间。随后打开建立的方法FA_ME_DB5MS_EI_V3_Scan,找到对应的组分并输入其保留时间,以此创建MRM方法FA_ME_MRM。随后将5×10^{-6},1×10^{-6},0.1×10^{-6},0.01×10^{-6}四个浓度标准品运行此方法,即可得到各组分标准曲线,后续可进行未知样品的绝对定量分析,也可以用此方法分析未知样品和对照样品进行相对定量分析。

时间的推移可能会造成保留时间的漂移,所以在间隔时间超过3个月后,需要再用1×10^{-6}标准品进行定标后更新方法文件来保证所测得样品各组分精准度。

2.30 脂肪酸分析

由于线虫产卵会消耗大量脂肪,且虫卵也富含的脂肪会对分析造成影响;同时 *let*-607 RNAi会影响线虫体型,而L4时期腹部有非常明显的白色半月型标志,所以本书针对野生型线虫N2进行control RNAi和 *let*-607 RNAi,于产卵前的L4时期幼虫萃取线虫游离脂肪酸,经过甲酯化处理后送重庆大学分析测试中心用GC-MS/MS进行脂质分析。实验过程中所用试剂均为色谱纯,所涉及容器均为玻璃或聚四氟乙烯材质。实验前处理方法参考美国著名脂类代谢科学家Jennifer Watts博士的文章[284],在不影响最终分析的情况下,结合本实验室硬件条件进行适当修改。

提前配置甲酯化反应液(含2.5%硫酸的甲醇溶液,当天配置当天使用),并且提前预热烘箱至80 ℃以便后续甲酯化反应。

用含0.01% Triton X-100的M9缓冲液将所需线虫(约20000只L4时期幼虫)从培养皿上洗到1.5 mL离心管中,静置使虫体沉淀到管底。用移液枪小心吸取M9缓冲液,随后用灭菌后的超纯水反复清洗虫体3次,尽可能清洗干净随虫体一起洗下的菌液，并且在最后一步尽可能吸净残留超纯水（水会干扰后续甲酯化反应）。

加入 1.5 mL 甲酯化反应液重悬虫体并将其转移到 10 mL 棕色玻璃样品瓶中，同时加入 10 μL C13 游离脂肪酸（5 mg/mL，Nu-chekprep，#N-13-A）作为内标物质，将样品瓶盖严（盖子使用 PTFE 硅胶垫圈），随即将样品瓶放置于提前预热好的 80 ℃烘箱中进行甲酯化反应 60 min，其间每 15 min 晃动一次样品瓶，促进甲酯化反应完全。

反应完后将样品瓶取出放置室温使其自然降温，切忌放入冰上快速降温，避免造成玻璃骤冷炸裂。待样品瓶降至室温后每个样品取出 0.1 mL 反应液用于总蛋白含量测定，再依次加入 2 mL 灭菌后的超纯水、1.5 mL 正己烷到样品瓶中，盖紧盖子用涡旋振荡器振荡 1 min，帮助正己烷萃取已经甲酯化完全的游离脂肪酸。随后将样品瓶用离心机以 3000 r/min 的速度离心 1 min，使正己烷完全浮于上层。再用 2 mL 注射器吸出上层正己烷，注意千万别吸到下层水相，最后使用 0.22 μm 有机相过滤器将正己烷过滤进 2 mL 棕色色谱进样瓶中，盖紧盖子置于冰上送重庆大学分析测试中心用 GC-MS/MS 检测。

待样品检测期间，立即对此前取出的 0.1 mL 反应液进行总蛋白含量测定，本书总蛋白使用 BCA 蛋白浓度测定试剂盒（碧云天，#P0012）测定。提前将全波长酶标仪（Molecular Devices，#SpectraMax 190）预热到 37 ℃。所有样品都需要设 3 个技术重复，取平均值进行计算。

将试剂盒中 BCA 试剂 A 和试剂 B 按照 50：1 的比例混匀，随后将试剂盒中 5 mg/mL BSA 标准品用 PBS 溶液稀释到 0，0.025，0.05，0.1，0.2，0.3，0.4，0.5 mg/mL 共 8 个梯度浓度，再在酶标板中加入 20 μL 各浓度蛋白标准品，随即加入 200 μL 配置好的 BCA 工作液。将酶标板置于已预热到 37 ℃的酶标仪中，用仪器振荡 15 s 以混匀反应液，随后孵育 30 min。设置酶标仪波长为 562 nm，待反应完全后检测各样品吸光值，为此建立一条标准曲线，其后可用此标准曲线为同时测定的未知样品进行绝对定量。

未知样品在不清楚其总蛋白浓度的情况下，需要用 PBS 等梯度稀释多个浓度进行测定，最后取吸光值最靠近标准蛋白样品 0.1~0.2 mg/mL 范围内吸光值的稀释浓度为最佳浓度。

2.31　薄层色谱法（TLC）

由于线虫产卵会消耗大量脂肪，且虫卵也富含的脂肪会对分析造成影响；同时 let-607 RNAi 会影响线虫体型，而 L4 时期腹部有非常明显的白色半月型标志，所以本书针对野生型线虫 N2 进行 control RNAi 和 let-607 RNAi，于产卵前的 L4 时期幼虫萃取线虫总脂质，经过薄层色谱法（thin-layer chromatography，TLC）分离线虫甘油三酯（triglyceride，TAG）和四种磷脂（Phospholipids，PLs）类脂质，再经过甲酯化处理后送重庆大学分析测试中心用 GC-MS/MS 进行脂质分析。实验过程中所用试剂均为色谱纯，所涉及容器均为玻璃或聚四氟乙烯材质。在不影响最终分析的情况下，结合本实验室硬件条件进行适当修改。此实验涉及挥发性有机物质需在通风橱内进行，并且某些不饱和脂质光照下不稳定需要避光操作。

提前准备如下有机溶剂和标准品：

氯仿（诺尔施）、甲醇（TEDIA，#MS1922-801），按照体积比 1∶1 配置 50 mL，室温保存；

展开液一：氯仿（诺尔施）、甲醇（TEDIA，#MS1922-801）、超纯水、乙酸（诺尔施），按照体积比 65∶43∶3∶2.5 配置 500 mL，室温保存；

展开液二：正己烷（TEDIA，#HS1722-801）、乙醚（成都科隆，科试）、乙酸（诺尔施），按照体积比 80∶20∶2 配置 500 mL，室温保存；

C13 游离脂肪酸（Nu-chekprep，#N-13-A），精确称取 10 mg 到 2 mL 棕色进样瓶中，加入 2 mL 氯仿配置工作浓度为 5 mg/mL 溶液，于 -20 ℃ 保存；

C13 甘油三酯（Nu-chekprep，#T-135），精确称取 11.374 mg 到 2 mL 棕色进样瓶中，加入 1 mL 氯仿配置浓度为 16.7 μmol/L 贮存溶液，再经过氯仿配置浓度为 16.7 nmol/L 工作溶液，两者皆于 -20 ℃ 保存；

C11 磷脂酰胆碱（Sigma-Aldrich，#850330P-25MG），直接加入 1.667 mL 氯仿至样品瓶中将其配置为 25 μmol/L 贮存溶液，再取出一部分用氯仿配置浓度为 25 nmol/L 工作溶液，两者皆于 -20 ℃ 保存；

L-α-磷脂酰胆碱（Sigma-Aldrich，#P3556-25MG,），直接加入 2.5 mL 氯仿至样品瓶中将其配置工作浓度为 10 mg/mL 溶液，于 -20 ℃ 保存；

3-sn-磷脂酰-丝氨酸（Sigma-Aldrich，#P7769-5MG），直接加入 0.5 mL 氯仿至样品瓶中将其配置工作浓度为 10 mg/mL 溶液，于 -20 ℃ 保存；

L-α-磷脂酰肌醇（Sigma-Aldrich，#P6636-250MG，10 mg/mL），精确称取10 mg标准品到2 mL棕色进样瓶中，加入1 mL氯仿配置工作浓度为10 mg/mL溶液，于-20 ℃保存；

L-α-磷脂酰乙醇胺（Sigma-Aldrich，#P7943-5MG），直接加入0.5 mL氯仿至样品瓶中将其配置工作浓度为10 mg/mL溶液，于-20 ℃保存；

0.005%樱草黄（生工生物，#A606348-0005），精确称取0.025 g樱草黄到500 mL喷壶，加入500 mL超纯水溶解后室温保存。

用含0.01% Triton X-100的M9缓冲液将所需线虫（约70000只L4时期幼虫）从培养皿上洗到1.5 mL离心管中，静置使虫体沉淀到管底。随后将虫体置于没有食物的干净培养皿上，这一步旨在使虫体消耗干净肠道内残留的大肠杆菌，避免这一部分细菌对实验最终结果的干扰。

约20分钟后用含0.01% Triton X-100的M9缓冲液将所需线虫从培养皿上洗到1.5 mL离心管中，静置使虫体沉淀到管底。用移液枪小心吸取M9缓冲液，随后用灭菌后的PBS缓冲液反复清洗虫体3次，并且在最后一步预留约500 μL液体。

提前将非接触式超声破碎仪（QSonica，#Q800R2）的水浴温度设为2 ℃（随超声过程水温会有上升，所以温度尽量低一些），将离心管置于仪器中，使水浴槽水位没过管内水位，调节仪器参数：超声工作10 s、停止20 s，功率70%，超声有效时间5 min。

超声完成后将离心管取出插到冰上，从其中吸出10 μL，用PBS等梯度稀释后测得其蛋白浓度。剩余部分全部转移至10 mL棕色样品瓶中，并加入20 μL C13_TAG（16.7 nmol/L）以及20 μL C11_PC（25 nmol/L），随即加入已提前预冷的1∶1氯仿甲醇溶液5 mL，在涡旋振荡器上使各组分充分混匀后将样品瓶置于-20 ℃静置过夜萃取。

第二天早上首先在硅胶薄层层析板上用铅笔标记出各组分对应位置，各组分点样位置应距离板左右下边缘2 cm，各样品的间隔也至少应为2 cm，并且在距离上边缘4.5 cm处也做一个标记，将标记好的硅胶薄层层析板置于玻璃层析缸内，一并放置于烘箱中（上海齐欣科学仪器，#DHG-9053A），并开机运行加热到110 ℃，处理1 h 15 min，这一步为活化硅胶薄层层析板并烘干水分。

在活化过程中将前一夜处理的样品取出，加入2.2 mL Hajra's溶液，在涡旋振荡器上振荡1 min，使1∶1氯仿甲醇溶液充分萃取各脂质组分。然后将样品瓶于3000 r/min的速度离心1 min使有机相和水相分层，用移液器小心吸取下层有机相到洁净的表面皿上。待有机相充分挥发后可看到表面皿上有白色脂质层，此时需用 1 mL 氯仿将析出脂质重新溶解后转移至 2 mL 棕色样品瓶，将所得样品置于冰上备用。

待硅胶薄层层析板活化时间到后取出，待其温度恢复至室温后即可上样，将标准品和位置样品依照此前标记好的位置依次用洁净的玻璃毛细管点到板上。在层析缸左槽加入适量展开液一（约 50 mL），将点好样的硅胶薄层层析板放到层析缸右槽，此过程为预饱和。

20 min 预饱和结束后，将层析缸左槽展开液一缓慢导入层析缸右槽，避光静置等待有机相从下往上展开。

大约 1 h 后展开液一会爬升至距离硅胶薄层层析板底部四分之三处（距层析板顶部 4.5 cm 处），随即取出硅胶薄层层析板，丢弃展开液一，换上展开液二再将硅胶薄层层析板放入。

大约 1 h 后展开液二爬升至硅胶薄层层析板顶端，立即取出硅胶薄层层析板，避光静置使板上有机相充分挥发后均匀喷上 0.005% 樱草黄，随后在紫外灯下对所分离脂质进行显影。

先找到各标准品在板上各自的位置，再用铅笔标记出未知样品中对应各标准品所在位置的脂质。而后使用洁净的载玻片将对应位置的硅胶薄层刮下，转移至 10 mL 棕色样品瓶中。加入 1.5 mL 甲酯化反应液重悬硅胶薄层，并加入 10 μL C13 游离脂肪酸（5 mg/mL，Nu-chekprep，#N-13-A）作为内标物质，将样品瓶盖严（盖子使用 PTFE 硅胶垫圈）。随即将样品瓶放置于提前预热好的 80 ℃ 烘箱中进行甲酯化反应 60 min，其间每 15 min 晃动一次样品瓶促进甲酯化反应完全。随后 GC-MS/MS 步骤同上。

2.32 秀丽隐杆线虫应激实验

本书如无特殊说明所有应激实验都在线虫 day 1 成虫期进行。

2.32.1 绿脓杆菌感染实验

配置秀丽隐杆线虫绿脓杆菌实验专用 NGM 平板：将普通平板组分中抗生素替换为工作浓度 25 μmol/L 的 5-氟-2′-脱氧尿嘧啶核苷（Sigma-Aldrich，#F0503-100MG）。

将同步化的 L1 时期幼虫转移至对应的实验用培养皿上，等待其长到 L4 时期。在使用无抗生素的LB液体培养基接种绿脓杆菌后于37 ℃ 220 r/min过夜培养，次日将5 μL菌液接种至绿脓杆菌实验专用NGM平板上，盖上培养皿盖子待菌液自然晾干后将此平板放到37 ℃的环境下培养12 h。

待线虫成长至L4幼虫期将其挑到绿脓杆菌培养皿上，后将培养皿置于25 ℃，每隔24 h计数一次死亡的线虫。

2.32.2 活性氧应激实验

将同步化的L1时期幼虫转移至对应的实验用培养皿上，等待其长到day 1时期。在进行活性氧应激实验前一夜将叔丁基过氧化氢（Sigma-Aldrich，#458139-25ML，10 mol/L）滴加到实验用培养皿上，使其工作浓度为10.0 mmol/L，盖上培养皿盖子使其自然晾干。于次日使用，从添加叔丁基过氧化氢到上虫不超过12 h。

待线虫成长至day 1期将其挑到添加了叔丁基过氧化氢的培养皿上，后将培养皿置于此前培养温度，每隔2 h计数一次死亡的线虫。

2.32.3 热应激实验

将同步化的L1时期幼虫转移至对应的实验用培养皿上，待线虫成长至day 1期将其挑到新的实验用培养皿上，后将培养皿置于35 ℃培养箱，每隔2 h计数一次死亡的线虫。

2.32.4 内质网应激实验

将同步化的L1时期幼虫转移至对应的实验用培养皿上，待线虫成长至day 1期。在进行内质网应激实验前一夜将二硫苏糖醇（Solarbio，#S8290-25 g）滴加到实验用培养皿上，使其工作浓度为8.5 mmol/L，盖上培养皿盖子使其自然晾干。于次日使用，从添加二硫苏糖醇到上虫不超过12 h。

待线虫成长至day 1期将其挑到添加了二硫苏糖醇的培养皿上，后将培养皿置于此前培养温度，每隔2 h计数一次死亡的线虫。

2.33 秀丽隐杆线虫寿命实验

配置秀丽隐杆线虫寿命实验专用NGM平板：将普通平板组分中抗生素替换为工作浓度25 μmol/L的5-氟-2'-脱氧尿嘧啶核苷（Sigma-Aldrich，#F0503-100MG）。

将同步化的L1时期幼虫转移至对应的普通实验用培养皿上,待线虫成长至L4幼虫期。在线虫进入L4时期前一天,将实验用菌液接种至寿命实验专用NGM平板上,待菌液在超净工作台中吹干后将此平板放到24 ℃环境中培养24 h。

待线虫成长至L4幼虫期将其挑到寿命实验培养皿上,后将培养皿置于20℃环境中,每隔2 d计数一次死亡的线虫。其中加了5-氟-2′-脱氧尿嘧啶核苷的寿命实验专用NGM平板需在L4时期用一次后,还应在day 2,day 4以及day 6各用一次,这样保证线虫在后续实验中不会出现如体内孵化等一系列与生殖相关的非正常死亡,同时也保证了产卵不孵化,避免下一代线虫对实验计数造成的影响。

2.34 秀丽隐杆线虫产卵量分析

本书分析了野生型秀丽隐杆线虫在不同生长时期进行 *let*-607 RNAi 后对产卵量的影响。首先将同步化处理的L1时期幼虫转移到control RNAi培养皿上,随后在L2时期和L4时期分别将培养在control RNAi培养皿上的线虫转移到 *let*-607 RNAi培养皿上,从而达到不同发育时期 *let*-607 RNAi的目的。

随后在各皿内线虫成长到L4时期时(该时期有明显的白色半月型标志,最容易区分同一时期),将线虫单独转移到新的RNAi皿上:将control RNAi培养皿内线虫转移至新的control RNAi培养皿内;L2 larva *let*-607 RNAi培养皿内线虫转移至新的L2 larva *let*-607 RNAi培养皿内;control RNAi培养皿内线虫转移至新的L4 larva *let*-607 RNAi培养皿内。每组准备15条线虫,每隔8 h换到新的皿上,一直计数到成虫后第四天。统计所有数据进行分析。

2.35 *let*-607p::GFP::*let*-607转基因线虫构建

由于 *let*-607 为人类 cAMP 反应元件结合蛋白3样3蛋白(cAMP responsive element binding protein 3 like 3,CREB3L3)直系同源物,编码环状单磷酸腺苷反应元件结合蛋白H(cAMP responsive element-binding protein H,CREBH),该蛋白在内质网合成后通过高尔基体上S1P(site-1 protease)和S2P(site-2 protease)两个蛋白酶,切割掉后半部分后,形成具有转录因子活性的蛋白。只有当有转录因子活性的CREBH进入细胞核后才能使其实现功能。所以本书将绿色荧光蛋白(green fluorescent protein,GFP)用同源重组的方法插入 *let*-607 的启动子和起始密码子之间。

2.35.1 构建 *let-607p::let-607* 载体

通过双酶切方法将 *let-607p::let-607* 片段连接到 pPD95_79 载体,正向引物选用内切酶 *Pst*I,保护碱基 AA,反向引物选用 *Acc*65I,保护碱基 GG:

*let-607p::let-607*_F: AA CTGCAG ACGTTGTTGACGACACTT

*let-607p::let-607*_R: GG GGTACC TAAAACATTTGGGTCTTTATTTTCTTAG

扩增片段为启动子 2143 bp + 5′UTR 216 bp + *let-607* 基因 4223 bp,共 6582 bp,将此片段连接到经 *Pst*I 和 *Acc*65I 双酶切后的 pPD95_79 载体,构建的 *let-607p::let-607* 载体为 11491 bp。

(1) *let-607p::let-607* 片段的扩增

设计的引物送生工生物(上海)有限公司合成,纯化方式为 PAGE,将合成的引物在 13000 r/min 的速度下离心 1 min 后用超纯水稀释为 5 mmol/L,再正反向引物等体积混合后得到 2.5 mmol/L 浓度的混合引物(反应体系最终引物浓度需在 0.2~0.3 μmol/L)。使用 PrimeSTAR® GXL DNA Polymerase(TaKaRa,#R050A)进行 PCR 扩增,反应体系如下:

组分	体积
5 × PrimeSTAR GXL Buffer	10 μL
dNTP Mixture(2.5 mmol/L)	4 μL
正反向混合引物(5 μmol/L)	5 μL
野生型线虫基因组 DNA	5 μL
PrimeSTAR GXL DNA Polymerase	1 μL
超纯水	补至 50 μL

将各组分添加到 200 μL 离心管中,涡旋振荡器混匀后置于 PCR 仪(Bio-Rad,T100™ Thermal Cycler,#1861096)运行程序:

模式	温度	时间	备注
预变性	98 ℃	2 min	
变性	98 ℃	10 s	
退火	62 ℃	15 s	
延伸	72 ℃	7 min	回到变性,循环 34 次
总延伸	72 ℃	10 min	
保存	16 ℃	∞	

提前配置0.7%琼脂糖凝胶，精确称取0.14 g琼脂糖到50 mL三角瓶中，加入20 mL TAE，微波炉加热至琼脂糖完全溶解，再在流水下冲洗三角瓶外壁帮助瓶内液体降温，待温度降至50 ℃左右时（置于手背部不明显感觉到烫）加入0.5 μL 4S Red Plus核酸染色剂（生工生物，#A606695-0100），晃动三角瓶使核酸染色剂充分混匀后倒入预先放置好的制胶槽中，大约15 min后温度降低使琼脂糖凝固后即可使用。

待程序运行完成后取出PCR产物，吸出2 μL到新的200 μL离心管中，加入7 μL超纯水，再加入1 μL 10X Loading Buffer，充分混匀后全部吸出转移至0.7%琼脂糖凝胶，再在旁边孔内加入5 μL 1Kb DNA Ladder（全式金，#BM201-02），使用100 V电压，280 mA电流运行25 min。随后置于凝胶成像系统（Tanon，2500）查看PCR产物是否为自己所需片段，如所得片段大小与设计一致，且无杂带则可对剩余样品进行PCR产物纯化。

（2）PCR产物纯化

PCR产物纯化使用生工生物（上海）有限公司产品SanPrep柱式PCR产物纯化试剂盒（#B518141-0100），具体操作步骤参见说明书。将得到的纯化后产物用50 μL超纯水洗脱，随后用Nanodrop 2000测得其浓度，-20 ℃冷冻保存。

（3）pPD95_79质粒扩增

将本实验室保存含有pPD95_79载体的菌接种到LB液体培养基中，于37 ℃ 220 r/min过夜培养，次日使用生工生物（上海）有限公司产品SanPrep柱式质粒DNA小量抽提试剂盒（B518191-0100），具体操作步骤参见说明书。将得到的质粒用50 μL超纯水洗脱，随后用Nanodrop 2000测得其浓度，-20 ℃冷冻保存。

（4）酶切

将所得PCR纯化产物使用 PstI（Thermo Scientific™，#ER0612）和 Acc65I（Thermo Scientific™，ER0901）双酶切系统在37 ℃恒温水浴酶切8 h，体系如下：

组分	体积
10X Buffer O	5 μL
PstI	1 μL
Acc65I	1 μL
PCR产物 / pPD95_79载体	1 μg
超纯水	补齐至50 μL

待酶切完成后，吸出2 μL到新的200 μL离心管中，加入7 μL超纯水，再加入1 μL 10X Loading Buffer，充分混匀后全部吸出转移至0.7%琼脂糖凝胶，再在旁边孔内加入5 μL 1Kb DNA Ladder（全式金，#BM201-02），使用100 V电压，280 mA电流运行25 min。随后置于凝胶成像系统查看酶切后的产物是否为自己所需片段，如所得片段大小与设计一致，且无杂带则可对剩余样品进行酶切产物纯化。

（5）酶切产物纯化

酶切产物纯化使用生工生物（上海）有限公司产品SanPrep柱式PCR产物纯化试剂盒（#B518141-0100），具体操作步骤参见说明书。将得到的纯化后产物用30 μL超纯水洗脱，随后用Nanodrop 2000测得其浓度，-20 ℃冷冻保存。

（6）连接

连接使用T4 DNA Ligase（Thermo Scientific™，#EL0011）进行，于1.5 mL离心管中加入如下体系：

组分	体积
10X T4 DNA Ligase Buffer	2 μL
pPD95_79载体酶切后纯化产物	50 ng
let-607p::let-607片段酶切后纯化产物	250 ng
T4 DNA Ligase	1 Weiss U
超纯水	补至20 μL

将1.5 mL离心管置于16 ℃连接过夜。

（7）转化大肠杆菌感受态细胞

从-80 ℃取出DH5α感受态细胞，插入冰上，同时从16 ℃将5 μL连接产物取出移至新的1.5 mL离心管也一并插入冰上，剩余连接产物于-20 ℃冻存备用。约5 min后，感受态细胞融化，用移液枪将100 μL感受态细胞移至5 μL连接产物离心管内，轻柔吹打混匀后立即插入冰上，同时准备一管阴性对照，即只用12.5 ng pPD95_79载体酶切后纯化产物转化100 μL感受态细胞。将两管插冰上保持15 min。

提前预热金属浴（杭州奥盛，MiniT-C）到42 ℃，待15 min后，将离心管置于42 ℃金属浴，热激90 s。热激结束后立即将离心管插入冰上，保持5 min。

5 min后可直接将转化细胞涂布到含100 μg/mL氨苄青霉素的LB固体培养基上，将培养皿倒置于37 ℃培养箱过夜培养。

（8）菌落PCR检测

转化第二日早上可见板上单菌落，配置含100 μg/mL氨苄青霉素的超纯水，取10 μL到200 μL PCR管中，随后用镊子夹取灭过菌的牙签沾取单菌落后再将牙签浸泡到含抗生素的超纯水中。晃动牙签取出丢弃，取2 μL菌液到新的PCR管中，用let-607p::let-607_F和let-607p::let-607_R进行PCR检测。剩余菌液接种到含100 μg/mL氨苄青霉素的液体LB培养基中，于37 ℃ 220 r/min培养8 h。

待检测结果出来后，选择阳性单克隆菌株保存后，再拷贝一份送生工生物（上海）有限公司测序。

2.35.2 构建let-607p::GFP::let-607载体

使用同源重组方式将GFP插入let-607起始密码子之前，设计引物如下所示：

```
           F1 →
Left Arm (let-607p)              Insert (GFP)                    F2 →
                                                          Right Arm (let-607)
5'- GTCGTCTTCCAT CGAAACGAAATG AGTAAAGGAGAAGAACTTTTCACTGGA···CATGGCATGGATGAACTATACAAA ATG GACCAAGATTTTGACCTCGAT GAA -3'
3'- CAGCAGAAGGTAGCTTTGCTTTAC TCATTTCCTCTTCTTGAAAAGTGACCT···GTACCGTACCTACTTGATATGTTT TAC TACCTGGTTCTAAAACTGGAGCTACTT -5'
                             ← R2                                        ← R1
```

（1）片段扩增和纯化

扩增插入片段，即GFP的引物，以pPD95_79质粒为模板进行扩增，引物序列如下。

F1：CGAAACGAAATGAGTAAAGGAGAAGAACTTTTCACT

R1：ATCGCGGTCAAAATCTTGGTCCATTTTGTATAGTTCATCCATGCC

扩增let-607p::let-607载体片段，用let-607p::let-607载体质粒为模板进行扩增，引物序列如下。

F2：GACCAAGATTTTGACCTCGAT

R2：TCCTTTACTCATTTCCTTTCGATGGAAGAC

使用PrimeSTAR® GXL DNA Polymerase（TaKaRa，#R050A）进行PCR扩增，对PCR产物使用生工生物（上海）有限公司产品SanPrep柱式PCR产物纯化试剂盒（#B518141-0100）进行纯化。将得到的纯化后产物用50 μL超纯水洗脱，随后用Nanodrop 2000测得其浓度，-20 ℃冷冻保存。

（2）片段连接

使用BG GeneMaster Pro试剂盒（葆光生物，BG0016）连接两个PCR产物片段，连接体系如下：

组分	体积
2.5X GM Buffer	4 μL
Em enhancer	1 μL
GFP 片段	125 ng
let-607p::let-607 线性载体	25 ng
超纯水	补齐至 10 μL

将各组分逐个添加到 200 μL PCR 管中，随后将其混匀后置于 PCR 仪中，于 50 ℃反应 30 min。

（3）转化大肠杆菌感受态细胞

从-80 ℃取出 BGT1 Ultracompetent Cells 感受态细胞，插入冰上；同时取 1 μL 连接产物移至新的 1.5 mL 离心管也一并插入冰上，剩余连接产物于-20 ℃冻存备用。约 5 min 后，感受态细胞融化，用移液枪将 50 μL 感受态细胞移至 1 μL 连接产物离心管内，吹打混匀后立即插入冰上；同时准备一管阴性对照，即只用 2.5 ng *let-607p::let-607* 线性载体转化 50 μL 感受态细胞。将两个离心管插冰上保持 4 min。

提前预热金属浴到 42 ℃，待冰浴 4 min 后，将离心管置于 42 ℃金属浴，热激 40 s。热激结束后立即加入 1 mL 无抗生素 LB 液体培养基，于 37 ℃ 200 r/min 培养 1 h。

1 h 后将离心管 3000 r/min 离心 1 min，弃 0.9 mL 上清液，将剩余 150 μL 菌体重悬后涂布到含 100 μg/mL 氨苄青霉素的 LB 固体培养基上，将培养皿倒置于 37 ℃培养箱培养 16 h。

（4）菌落 PCR 检测

转化第二日早上可见板上单菌落，配置含 100 μg/mL 氨苄青霉素的超纯水，取 10 μL 到 200 μL PCR 管中，随后用镊子夹取灭过菌的牙签沾取单菌落后再将牙签浸泡到含抗生素的超纯水中。晃动牙签取出丢弃，取 2 μL 菌液到新的 PCR 管中，用 *let-607p::let-607*_F 和 R1 进行 PCR 检测，如果成功插入 GFP 片段，则 PCR 产物扩增片段为启动子 2143 bp + 5′ UTR 216 bp + GFP 870 bp = 3229 bp；未插入片段扩增片段为启动子 2143 bp + 5′ UTR 216 bp = 2359 bp，多出的 870 bp 通过琼脂糖凝胶电泳会很明显区分开来。剩余菌液接种到含 100 μg/mL 氨苄青霉素的液体 LB 培养基中，于 37 ℃ 220 r/min 培养 8 h。

待检测结果出来后，选择阳性单克隆菌株保存一份后，再拷贝一份送生工生物（上海）有限公司测序。

3.35.3 显微注射

对野生型秀丽隐杆线虫进行 let-607p::GFP::let-607 载体的显微注射，以构建 LET-607::GFP 转基因线虫。

将注射混合液按照以下浓度要求配置 20 μL，混匀后以 8000 r/min 的速度离心 5 min 后吸出上层的 10 μL 用于注射。

组分	浓度
let-607p::GFP::let-607 载体	7.5 ng/μL
myo-2p::mcherry	25 ng/μL
	共配置 20 μL

具体显微注射操作步骤如下：

首先取出一支新的注射玻璃针管（NARISHIGE，#GD-1）到拉针器（NARISHIGE，#PC-10）上固定好，经过多次摸索，经过 66.6 ℃加热后拉出的针最适合用于显微注射；

用移液器将注射液吸入到显微注射针中，后将针装载到显微注射气泵（NARISHIGE，#IM-31）上，通入 99.5%纯度的工业氮气，用 120 kPa 进行显微注射；

每次注射超过 20 只 day 1 线虫，随后通过荧光观察其子代，如果观察到子代线虫咽部有红色荧光，则代表显微注射成功，目的质粒表达。

2.35.4 整合 let-607p::GFP::let-607 载体到线虫基因组

挑取大约 120 只 L4 时期带有红色荧光的线虫到空的 NGM 皿上，放到 Crosslinker（analytic jena，#UVlink 1000）中，用 12.5 mJ/cm^2 的 UV 照射线虫，随后将处理过的线虫转移到 OP50-1 的 NGM 皿上，转移至 25 ℃培养。

约 4 天之后，挑出约 200 只带有红色荧光的 L4 时期子代 F1 到 OP50-1 的 NGM 皿上，每皿 1 只。

再过约 4 天之后，找到 F2 代中带有红色荧光比例最多的一皿，挑出约 200 只带有红色荧光的 L4 时期子代 F2 到 OP50-1 的 NGM 皿上，每皿 1 只。

再过约 4 天之后，找到全部带有红色荧光 F3 代的皿，挑出约 20 只带有红色荧光的 L4 时期子代 F3 到 OP50-1 的 NGM 皿上，每皿 1 只。

再过约4天之后，观察此前挑出20皿中的F4是否全有红色荧光，若全有，则取出部分在正置激光共聚焦显微镜下观察是否全有LET-607::GFP的绿色荧光表达，若全有，则整合成功，随后使用雄性野生型秀丽隐杆线虫和整合后的线虫进行多代杂交，以替换掉除整合到基因组的 let-607p::GFP::let-607 外的其他位置，以保证其他位置的基因没有因为紫外的照射而产生突变，得到野生型背景的LET-607::GFP转基因线虫。

2.36 化合物补充实验

2.36.1 神经酰胺

将各种类神经酰胺 C16-ceramide（Sigma-Aldrich，#860516P-5mg），C18-ceramide（Sigma-Aldrich，#860517P-5mg），C20-ceramide（Sigma-Aldrich，#860520P-5mg），C22-ceramide（Sigma-Aldrich，#860501P-5mg），C24-ceramide（Sigma-Aldrich，#860524P-5mg）溶解在乙醇中，制成0.5 mg/mL的贮存溶液，于-20 ℃保存。

将同步化处理后的L1时期实验用线虫转移RNAi菌培养皿中，置于所需温度的培养箱培养。在线虫成长到L4时期前一天，重新准备RNAi菌接种到NGM培养皿上，于25 ℃诱导24 h后，取各神经酰胺25 μg均匀接种到菌斑表面，盖上培养皿盖子在超净工作台中使液体蒸干，随后将L4时期的线虫转移到培养皿中，于第二日早上观察并拍照。

2.36.2 磷脂酰胆碱

将各种类磷脂酰胆碱 DPPC（16:0 PC，Sigma-Aldrich，#850355P-25MG），DSPC（18:0 PC，Sigma-Aldrich，#850365P-25MG），DOPC（18:1 PC，Sigma-Aldrich，#P6354-25MG），DLPC（18:2 PC，Sigma-Aldrich，#850385C-25MG）溶解在DMSO中，制成17 mmol/L贮存溶液，于-20 ℃保存。

将同步化处理后的L1时期实验用线虫转移RNAi菌培养皿中，置于所需温度的培养箱培养。在线虫成长到L4时期前一天，重新准备RNAi菌接种到NGM培养皿上，于25 ℃诱导24 h后，取各磷脂酰胆碱51 nmol/L均匀接种到菌斑表面，盖上培养皿盖子在超净工作台上使液体蒸干，随后将L4时期的线虫转移到培养皿中，于第二日早上观察并拍照。

2.36.3 离子霉素

将离子霉素（ionomycin，Sigma-Aldrich，#407951-1MG）溶解在 DMSO 中，制成 0.705 mmol/L 贮存溶液于 $-20\ ℃$ 保存。

将同步化处理后的 L1 时期实验用线虫转移至 RNAi 菌培养皿中，置于所需温度的培养箱培养。在线虫成长到 L4 时期前一天，重新准备 RNAi 菌接种到 NGM 培养皿上，于 25 ℃ 诱导 24 h 后，取 0.705 mmol/L 贮存溶液 2.84 μL，加 50 μL 灭菌后超纯水稀释后均匀接种到的菌斑表面，使其工作浓度为 5 μmol/L，盖上培养皿盖子在超净工作台上使液体蒸干，随后将 L4 时期的线虫转移到培养皿中，于第二日早上观察并拍照。

2.36.4 1-油酰基-2-乙酰基-sn-甘油

将 1-油酰基-2-乙酰基-sn-甘油（18∶1-2∶0 DG，Sigma-Aldrich，#800100O-10MG）溶解在 DMSO 中，制成 2 mg/mL 贮存溶液于 $-20\ ℃$ 保存。

将同步化处理后的 L1 时期实验用线虫转移 RNAi 菌培养皿中，置于所需温度的培养箱培养。在线虫成长到 L4 时期前一天，重新准备 RNAi 菌接种到 NGM 培养皿上，于 25 ℃ 诱导 24 h 后，取 2 mg/mL 贮存溶液 6 μL，加 50 μL 灭菌后超纯水稀释后均匀接种到菌斑表面，使其工作浓度为 3 μg/mL，盖上培养皿盖子在超净工作台中使液体蒸干，随后将 L4 时期的线虫转移到培养皿中，于第二日早上观察并拍照。

2.36.5 氯化胆碱

将氯化胆碱（choline chloride，Sigma-Aldrich，#C7017-10MG）溶解在超纯水中，制成 1 mol/L 贮存溶液于 $-20\ ℃$ 保存。

将同步化处理后的 L1 时期实验用线虫转移至 RNAi 菌培养皿中，置于所需温度的培养箱培养。在线虫成长到 L4 时期前一天，重新准备 RNAi 菌接种到 NGM 培养皿上，于 25 ℃ 诱导 24 h 后，取 1 mol/L 贮存溶液经稀释后均匀接种到菌斑表面，使其工作浓度为 50 μmol/L，盖上培养皿盖子在超净工作台中使液体蒸干，随后将 L4 时期的线虫转移到培养皿中，于第二日早上观察并拍照。

2.36.6 地昔帕明

将地昔帕明（desipramine，Sigma-Aldrich，#D3900-1G）溶解在 DMSO 中，

制成100 mmol/L贮存溶液于-20 ℃保存。

将同步化处理后的L1时期实验用线虫转移至RNAi菌培养皿中，置于所需温度的培养箱培养。在线虫成长到L4时期前一天，重新准备RNAi菌接种到NGM培养皿上，于25 ℃诱导24 h后，取100 mmol/L贮存溶液经稀释后均匀接种到菌斑表面，使其工作浓度为50 μmol/L，盖上培养皿盖子在超净工作台上使液体蒸干，随后将L4时期的线虫转移到培养皿中，于第二日早上观察并拍照。

2.37 数据量化和统计分析

数据表示为平均值±SEM；使用log-rank（Mantel-Cox）test分析生存数据；使用Chi-square and Fisher's exact test分析DAF-16::GFP的核积累；0.771650小于使用unpaired student t-test分析单因素变量荧光半定量数据；使用multiple t-test followed by a Holm-Sidak post hoc test分析GC-MS/MS数据；使用One-way ANOVA分析MTL-1::GFP的 *ire*-1，*pek*-1，*atf*-6 RNAi荧光半定量数据；使用Two-way ANOVA followed by a Tukey post hoc test分析多因素变量荧光半定量数据；使用online software provided by Jim Lund 分析 *let*-607 RNAi和DAF-16 I类靶基因诱导的重叠基因数。$P<0.05$，标识为*；$P<0.01$，标识为**；$P<0.001$，标识为***。

分析统计结果见附录部分：附录F.3为病原菌感染实验统计结果，附录F.4为活性氧应激实验统计结果，附录F.5为热应激实验统计结果，附录F.6为内质网应激实验统计结果，附录F.7为寿命实验统计结果。

2.38 主要仪器

本书使用的主要仪器有：离心机（湘仪，L420），混匀仪（大龙，MX-RL-E），冷冻离心机（Sigma，3K15），凝胶成像系统（Tanon，2500），灌装蠕动泵（兰格，WT600-1F），恒温培养箱（宁波江南，SPX-50），迷你金属浴（杭州奥盛，MiniT-C），恒温培养箱（上海博讯，BSP-400），超净工作台（苏州净化，SW-CJ-1D），恒温培养摇床（上海一恒，THZ-100），高压灭菌锅（上海博讯，YXQ-LS-50A），恒温水浴锅（科析仪器，KW-1000DC），琼脂糖凝胶电泳仪（北京六一，DYY-7C），-80 ℃超低温冰箱（澳柯玛，DW-

86L500），万分之一天平（Mettler Toledo，AL204-IC），微型台式真空泵（海门其林贝尔，GL-802A），荧光连续变倍体视显微镜（Nikon，SMZ1270），连续变倍体视显微镜（奥特光学，SZ680B2L），正置激光共聚焦显微镜（Leica，TCS SP8），三重四极杆型气相色谱质谱联用仪（岛津，GCMS-TQ8040），色谱柱（Shimadzu，SH-Rxi-5sil，#221-75954-30，30 m × 0.25 mm），拉针器（NARISHIGE，#PC-10），显微注射气泵（NARISHIGE，#IM-31）。

3 结　果

3.1 LET-607对秀丽隐杆线虫应激反应的影响

3.1.1 LET-607对线虫先天免疫的影响

本实验室前期研究结果显示，将幼虫时期秀丽隐杆线虫暴露于较高温度下，线虫先天免疫的持续激活可增强其抵抗应激的能力并且延长其寿命[268]。为了找到参与反应的调控因子，本书利用成熟的免疫报告基因 *T24B8.5* 的转基因线虫 AU78（*T24B8.5p*::GFP）对秀丽隐杆线虫转录因子文库进行了RNAi筛选[285]。结果发现，将培养在25 ℃环境下AU78的 *let*-607 RNAi后，观察到其GFP较对照组明显变暗（见图3.1）。

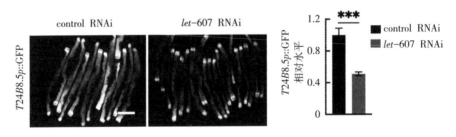

图3.1　*let*-607 RNAi对 *T24B8.5p*::GFP的影响

直接用 *let*-607 RNAi细菌培养物喂食会导致秀丽隐杆线虫的发育停滞，因此本书使用control RNAi（L4440空载体转化HT115感受态细胞）稀释后的 *let*-607 RNAi（control RNAi:*let*-607 RNAi＝4∶1）确保线虫发育为成年动物，同时对野生型秀丽隐杆线虫N2进行 *let*-607 RNAi处理后，通过QPCR检测，其有效地使 *let*-607 mRNA水平降低约75%［见图3.2（a）］，并且也检测到 *T24B8.5* 的mRNA水平显著降低［见图3.2（b）］。

3 结果

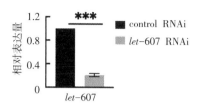
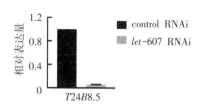

（a） *let*-607 RNAi 对 *let*-607 mRNA 表达的影响

（b） *let*-607 RNAi 对 *T24B*8.5 mRNA 表达的影响

图 3.2 *let*-607 RNAi 对 *let*-607 和 *T24B*8.5 mRNA 表达的影响

保守的 PMK-1/p38 丝裂原活化蛋白激酶是秀丽隐杆线虫先天免疫防御系统的主要组成部分，该系统调节先天免疫和解毒基因的表达[269,286]。因此，我们通过蛋白质免疫印迹（Western Blot）验证 PMK-1 是否介导 *let*-607 RNAi 调控的防御反应。结果显示磷酸化 PMK-1 蛋白（p-PMK-1）也显著降低（见图 3.3），与如图 3.1 所示荧光结果一致。

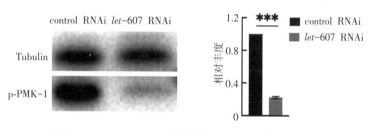

图 3.3 *let*-607 RNAi 对磷酸化 PMK-1 蛋白的影响

但有趣的是，进一步使用绿脓杆菌（*p.aeruginosa*，PA）检测线虫对病原体抗性，结果显示 *let*-607 RNAi 反而增强了野生型线虫 N2 对 PA 的抗性（见图 3.4）。

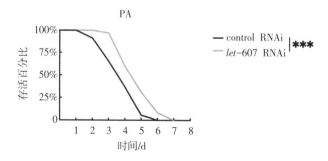

图 3.4 *let*-607 RNAi 对野生型线虫病原体抗性的影响

57

以上实验结果说明，LET-607在线虫免疫应答中起着关键作用，但是其增强线虫抗病性却不直接依赖*T24B*8.5。

前期的研究结果表明，秀丽隐杆线虫对病原体的抗性有包括DAF-16[269]、HSF-1[287]和SKN-1[288-289]在内的几种胞质转录因子的参与，而且此前也有报道揭示*let*-607 RNAi促进了HSF-1靶基因的表达[242]。由此推测*let*-607 RNAi可能激活上述转录因子来增强线虫的抗逆性，我们对携带这些转录因子各自报告基因的转基因线虫进行*let*-607 RNAi。结果表明*let*-607 RNAi增强了DAF-16报告基因MTL-1::GFP和*sod*-3*p*::GFP（见图3.5），以及HSF-1报告基因*hsp*-16.2*p*::GFP[290]和*hsp*-70*p*::GFP（见图3.6），但是SKN-1报告基因*gst*-4*p*::GFP却不受影响（见图3.7）[291]。

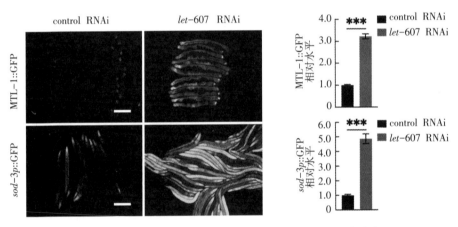

图3.5 *let*-607 RNAi对MTL-1::GFP和*sod*-3*p*::GFP的影响

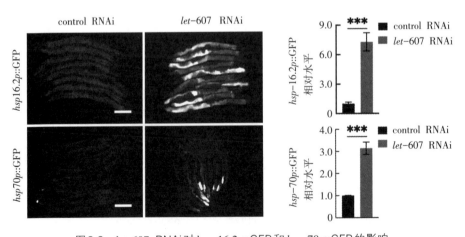

图3.6 *let*-607 RNAi对*hsp*-16.2*p*::GFP和*hsp*-70*p*::GFP的影响

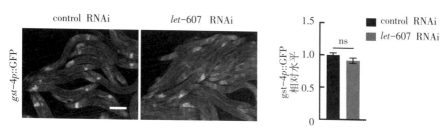

图3.7 *let*-607 RNAi对*gst*-4*p*::GFP的影响

进一步对突变体线虫*daf*-16（*mu86*）和*hsf*-1（*sy441*）线虫进行*let*-607 RNAi后分别进行PA实验。实验结果表明，在*daf*-16缺失突变后，*let*-607 RNAi后增强PA存活率的性状消失了；而*hsf*-1的缺失突变却没有影响（见图3.8）。

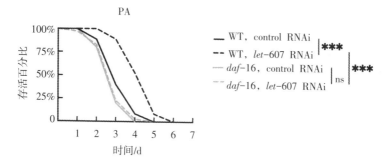

（a）*let*-607 RNAi对*daf*-16突变型线虫病原体抗性的影响

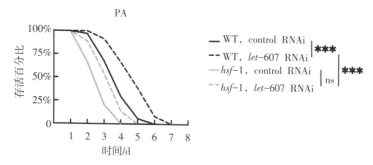

（b）*let*-607 RNAi对*hsf*-1突变型线虫病原体抗性的影响

图3.8 *let*-607 RNAi对*daf*-16和*hsf*-1突变型线虫病原体抗性的影响

以上实验结果说明，*let*-607可以激活*hsf*-1及*daf*-16下游的基因，并且在线虫免疫应答过程中LET-607增强线虫抗病性依赖DAF-16。

3.1.2 LET-607对线虫抵抗外界压力的影响

前面实验结果已经证实,在线虫免疫应答过程中LET-607增强线虫抗病性依赖DAF-16。接下来我们进一步验证LET-607是否参与到包括叔丁基过氧化氢诱导的活性氧应激、35℃热应激以及二硫苏糖醇诱导的内质网应激等应激反应中。实验结果表明,let-607的RNAi对包括氧化应激[见图3.9(a)]、热应激[见图3.9(b)]、内质网应激[见图3.9(c)]在内的应激反应都有所增强。由于抗逆性有助于寿命的延长,在对野生型线虫let-607 RNAi后,也观察到其寿命显著增加[见图3.9(d)]。

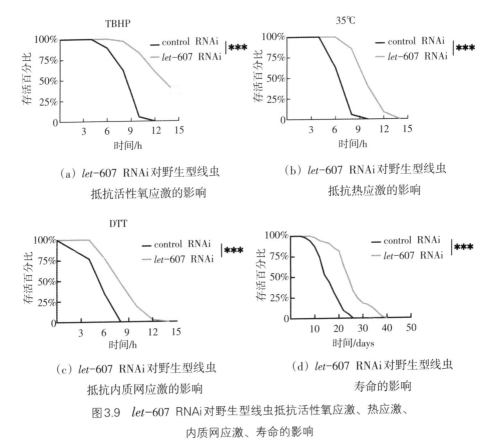

(a) let-607 RNAi对野生型线虫抵抗活性氧应激的影响

(b) let-607 RNAi对野生型线虫抵抗热应激的影响

(c) let-607 RNAi对野生型线虫抵抗内质网应激的影响

(d) let-607 RNAi对野生型线虫寿命的影响

图3.9 let-607 RNAi对野生型线虫抵抗活性氧应激、热应激、内质网应激、寿命的影响

进一步通过组织特异性RNAi的转基因线虫进行热应激和氧化应激,结果显示只有肠道内的let-607 RNAi才会增强线虫的热激和氧化应激反应,在性腺、肌肉和表皮都没有作用(见图3.10)。

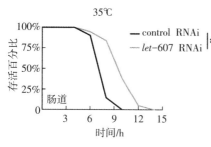

(a) *let*-607 RNAi对肠道特异性RNAi
线虫抵抗热应激反应的影响

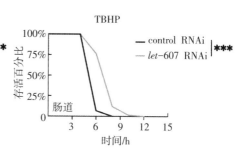

(b) *let*-607 RNAi对肠道特异性RNAi
线虫抵抗氧化应激反应的影响

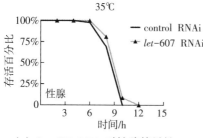

(c) *let*-607 RNAi对性腺特异性RNAi
线虫抵抗热应激反应的影响

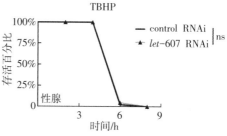

(d) *let*-607 RNAi对性腺特异性RNAi
线虫抵抗氧化应激反应的影响

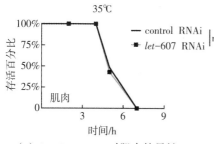

(e) *let*-607 RNAi对肌肉特异性RNAi
线虫抵抗热应激反应的影响

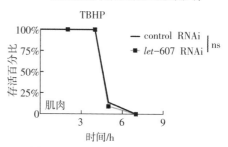

(f) *let*-607 RNAi对肌肉特异性RNAi
线虫抵抗氧化应激反应的影响

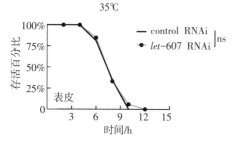

(g) *let*-607 RNAi对表皮特异性RNAi
线虫抵抗热应激反应的影响

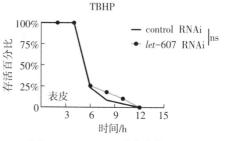

(h) *let*-607 RNAi对表皮特异性RNAi
线虫抵抗氧化应激反应的影响

图3.10 *let*-607 RNAi对组织特异性RNAi线虫抵抗热应激和氧化应激反应的影响

我们进一步检查了 let-607 RNAi 对线虫生殖发育的影响，发现后代数量大大减少 [见图3.11（a）]，这一结果也与先前的报道一致[244]。并且，对L4时期线虫体长的检测也发现 let-607 RNAi 对线虫体型影响也是巨大的 [见图3.11（b）]。

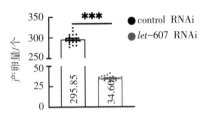
（a） let-607 RNAi 对野生型线虫后代数量的影响

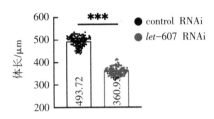
（b） let-607 RNAi 对野生型线虫L4时期体长的影响

图3.11　let-607 RNAi 对野生型线虫后代数量和L4时期体长的影响

由于 let-607 RNAi 会影响线虫发育，因此本书还对发育后的线虫进行了 let-607 RNAi 实验。结果表明，对发育后的线虫进行 let-607 RNAi，发现在产卵量 [见图3.12（a）] 和体长 [见图3.12（b）] 都与对照组无差异的情况下，同样可以增强其氧化应激能力 [见图3.13（a）]、热应激能力 [见图3.13（b）] 和寿命 [见图3.13（c）]。LET-607 在发育和寿命中的作用可能说明其具有基因多效性。

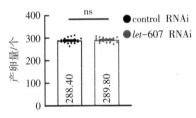
（a） let-607 RNAi 对发育后野生型线虫后代数量的影响

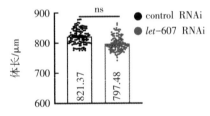
（b） let-607 RNAi 对发育后野生型线虫day 2时期体长的影响

图3.12　let-607 RNAi 对发育后野生型线虫后代数量和day 2时期体长的影响

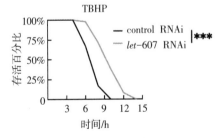
（a） let-607 RNAi 对发育后野生型线虫抵抗活性氧应激的影响

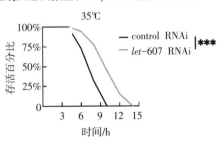
（b） let-607 RNAi 对发育后野生型线虫抵抗活性热应激的影响

3 结果

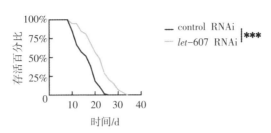

（c）let-607 RNAi对发育后野生型线虫抵抗活性寿命的影响

图3.13 let-607 RNAi对发育后野生型线虫抵抗活性的影响

这些数据表明LET-607负调节秀丽隐杆线虫的抗逆性和寿命。由于抗逆性通常由多种逆境反应途径的最终结果决定，因此我们推测let-607 RNAi可能会自适应激活其他途径，从而增强抗逆性和延长寿命。

3.1.3 秀丽隐杆线虫LET-607的细胞定位

此前多人研究结果显示let-607与人类crebh同源，而人源CREBH为内质网锚定蛋白。为了了解秀丽隐杆线虫LET-607在细胞中的定位，我们构建了let-607p::gfp::let-607转基因线虫以进行观察。DIS-3是细胞核RNA外泌体复合物的一个组成部分，在核质中富集，但在核仁中却不富集[292]。将LET-607::GFP和DIS-3::mCHERRY线虫杂交后，使用正置激光共聚焦系统观察，确定LET-607::GFP在核质中与DIS-3::mCHERRY重叠（见图3.14），表明LET-607的核定位到LET-607在秀丽隐杆线虫细胞中定位在细胞核而非内质网。

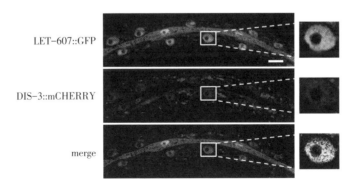

图3.14 LET-607::GFP表达在核质

为此著者比对分析了LET-607和CREBH的氨基酸序列，发现在影响到内质网锚定的跨膜结构域两者有较大的不同（见图3.15），这可能就是LET-607定位在细胞核的主要原因，并且此前关于果蝇CREBH的研究也表明其因为缺少跨膜结构域而没有锚定到内质网[293]。

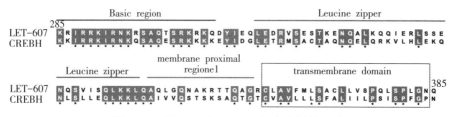

图3.15 LET-607与CREBH蛋白序列的比对

同时通过对LET-607::GFP线虫进行 let-607 RNAi也观察到其入核减少（见图3.16），此结果也与QPCR（见图3.2）结果一致。

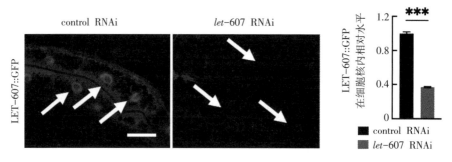

图3.16 let-607 RNAi对LET-607::GFP的影响

本节主要发现秀丽隐杆线虫LET-607作为人类CREBH的同源蛋白，定位于核质而不是内质网中。LET-607负调控秀丽隐杆线虫包括对致病菌、活性氧应激、热应激、内质网应激在内的应激反应以及寿命，并且通过活性氧应激和热应激实验发现当肠道内特异的 let-607RNAi才有这一增强作用；同时，发现 let-607RNAi还会影响线虫正常的生殖和发育，但在线虫发育完成后再进行RNAi，LET-607对线虫的应激抗性增强和寿命延长现象依旧存在。本节发现了一个新的长寿调控因子。

3.2 LET-607通过DAF-16调控线虫抗压能力和寿命

3.2.1 LET-607对基因表达的影响

为了剖析LET-607是通过何种方式参与增强线虫抗逆性和寿命的，著者收集了L4时期线虫的mRNA进行了转录组测序及分析。结果显示，LET-607

调控了860个基因,包括316个基因被上调,544个基因被下调。基因功能分类分析揭示了LET-607上调了大量对病原体防御重要的C型凝集素和先天免疫基因,这与哺乳动物CREBH的功能一致;同时,对细胞蛋白稳态至关重要的肽酶和介导解毒反应的细胞色素P450也被上调表达(见图3.17)。通过这些结果可以得知LET-607积极调节参与多种应激反应的基因,主要是防御反应和蛋白稳态控制反应,这表明某些LET-607功能与哺乳动物CREBH保持一致。

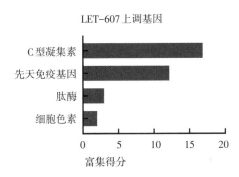

图3.17 *let*-607 RNAi上调基因的功能分类分析

但有趣的是,经典的内质网未折叠蛋白反应报告基因 *hsp*-3,*hsp*-4 不受 *let*-607 RNAi 的调控(见图3.18);由于LET-607/CREBH可以响应ER应激[65,267],而通过转基因线虫SJ4005(*hsp*-4::GFP)观察到 *let*-607 RNAi 抑制了衣霉素诱导的内质网应激反应(见图3.19),这也符合此前报道的LET-607可以结合到 *hsp*-4 的启动子区域[244]。

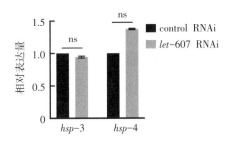

图3.18 *let*-607 RNAi对 *hsp*-3和 *hsp*-4 mRNA表达量影响

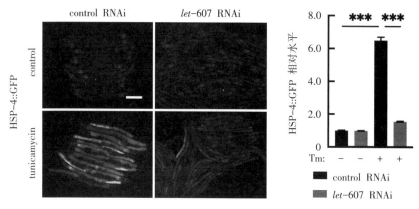

图3.19 衣霉素处理下 let-607 RNAi 对 HSP-4::GFP 的影响

因此，我们进一步测试了抑制经典 ER 应激反应途径是否也可以激活 DAF-16。通过对 MTL-1::GFP 转基因线虫进行 ER^{UPR} 基因（ire-1，pek-1 和 atf-6）的 RNAi，发现其荧光的表达没有影响（见图3.20），这些基因和它们的靶基因如预期的那样都被下调表达了（见图3.21）。这些数据表明，LET-607 独立于内质网未折叠蛋白反应而激活 DAF-16。

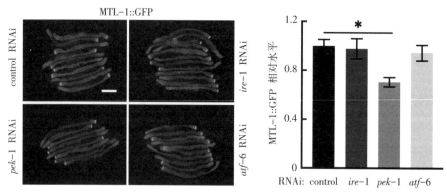

图3.20 ER UPR 基因 RNAi 对 MTL-1::GFP 的影响

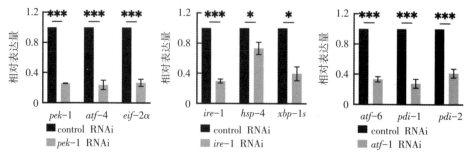

图3.21 ER^{UPR} 基因 RNAi 后对其自身与下游靶基因表达量的影响

3.2.2 LET-607通过调节DAF-16入核来激活其功能

通过QPCR验证可知，DAF-16另一个报告基因 sod-3 的mRNA也上调表达了（见图3.22），这与其GFP变化结果一致（见图3.5）。值得注意的是 daf-16 的mRNA水平却不受 let-607 RNAi的影响（见图3.22）。

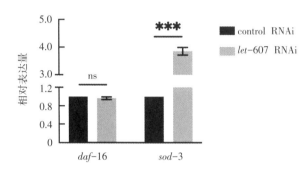

图3.22　let-607 RNAi对 sod-3 和 daf-16 mRNA表达量的影响

因为DAF-16被激活后会转移至细胞核，所以我们观察到在转基因线虫TJ356（DAF-16::GFP）let-607 RNAi处理后，DAF-16入核比例增加了（见图3.23），由此可以推测，LET-607通过调节DAF-16的入核来激活其功能。

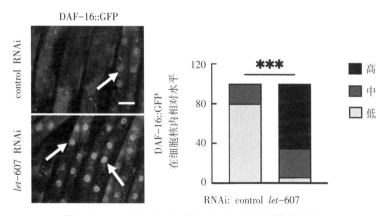

图3.23　let-607 RNAi对DAF-16::GFP入核的影响

以上结果表明，LET-607通过负调控细胞质应激反应因子DAF-16入核来激活其行使功能。

基于LET-607在线虫免疫应答中增强线虫抗病性依赖DAF-16，由此推测 let-607 RNAi也可能激活DAF-16依赖性应激反应基因。此前的研究结果将

DAF-16应答基因分为两类，分别为1633个DAF-16阳性靶基因（class Ⅰ）和1733个DAF-16阴性靶基因（class Ⅱ）[294]。著者将 *let*-607 RNAi诱导上调表达的基因与DAF-16阳性靶基因进行比较，发现有74个基因（13.6%），结果大大超过偶然比例（见图3.24），这进一步支持 *let*-607 RNAi调节了DAF-16的活性。

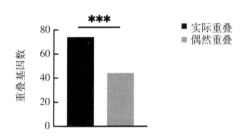

图3.24 *let*-607 RNAi和DAF-16 Ⅰ类靶基因诱导的重叠基因数

接下来，我们对这些重叠基因进行生物学过程中的基因本位论（gene ontology，GO）分析，结果显示这些基因在多个类别中富集，包括多重应激反应、氧化还原过程和寿命相关过程（见图3.25），这个结果与 *let*-607 RNAi后诱导的表型一致。

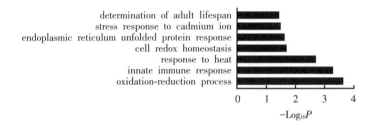

图3.25 *let*-607 RNAi 诱导的DAF-16 Ⅰ类靶基因的GO分析

2016年Weicksel等人已经通过ChIP-seq分析鉴定了在启动子区域内具有LET-607结合位点的基因。将这些ChIP-seq的原始数据分析后得到的1133个call peak，进一步将 *let*-607 RNAi上调的DAF-16靶基因与这些进行比较，发现只有 *sodh*-1这一个基因包含LET-607结合位点，表明LET-607不能直接调控这些基因，很可能是通过DAF-16起作用。

综上所述，*let*-607负调控DAF-16，并且DAF-16下游应激反应基因的激活可能有助于解释 *let*-607 RNAi线虫增强的应激抗性。

3.2.3 LET-607通过DAF-16调节抗压力

接下来，验证DAF-16激活是否决定let-607 RNAi处理的线虫抗逆性的增强。对突变体线虫daf-16（mu86）和hsf-1（sy441）线虫let-607基因 RNAi后进行活性氧应激和热应激实验。结果显示，daf-16的缺失部分抑制了线虫对活性氧的抵抗能力［见图3.26（a）］，而不会对热应激有明显影响［见图3.26（b）］。而hsf-1的缺失则部分抑制了线虫对热应激的抵抗能力［见图3.26（c）］，却对活性氧不敏感［见图3.26（d）］。

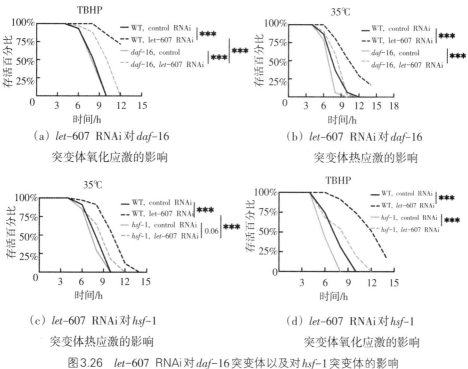

（a）let-607 RNAi对daf-16
突变体氧化应激的影响

（b）let-607 RNAi对daf-16
突变体热应激的影响

（c）let-607 RNAi对hsf-1
突变体热应激的影响

（d）let-607 RNAi对hsf-1
突变体氧化应激的影响

图3.26 let-607 RNAi对daf-16突变体以及对hsf-1突变体的影响

综合上述结果可以得出结论：DAF-16主要负责提高let-607 RNAi线虫的防御反应，HSF-1则负责维持蛋白稳态，并支持其他应激反应途径的适应性激活，有助于增强抗逆性。

3.2.4 LET-607激活DAF-16有助于增加线虫寿命

由于DAF-16是寿命的重要调控因子，因此进一步研究DAF-16的诱导是否有助于线虫寿命的增加。首先使用突变体线虫daf-16（mu86）进行验证，如预期的那样，*daf*-16 的缺失在很大程度上消除了 *let*-607 RNAi 而延长了线虫寿命（见图 3.27）。

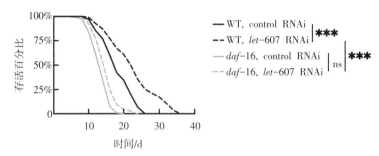

图 3.27　*let*-607 RNAi 对 *daf*-16 突变体寿命的影响

此前的研究报道胰岛素/IGF 受体 *daf*-2 [141-142,149] 和 *glp*-1 [135,295] 是 DAF-16 延长线虫寿命所必需的。著者发现 *let*-607 RNAi 可以进一步延长 *daf*-2 突变体的寿命［见图 3.28（a）］，而未能影响 *glp*-1 突变体的寿命［见图 3.28（b）］。这表明 *let*-607 RNAi 和生殖细胞缺失突变体缺陷模型之间可能存在共享的寿命延长机制。

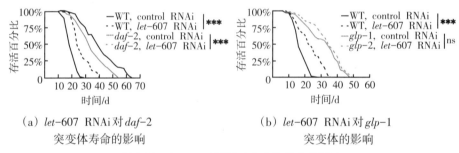

（a）*let*-607 RNAi 对 *daf*-2
　　突变体寿命的影响

（b）*let*-607 RNAi 对 *glp*-1
　　突变体的影响

图 3.28　*let*-607 RNAi 对突变体寿命的影响

著者统计到 *let*-607 RNAi 后线虫后代数量大大减少，如图 3.11（a）所示，但是对于延长生殖细胞缺失突变体、缺陷型突变体的寿命至关重要的基因，包括 *daf*-9 和 *daf*-12 [296-297]，*tcer*-1 [298-299] 以及 *kri*-1 [194,295]，对于 *let*-607 RNAi 后线虫的寿命延长是必不可少的［见图 3.29（a）~图 3.29（d）］。

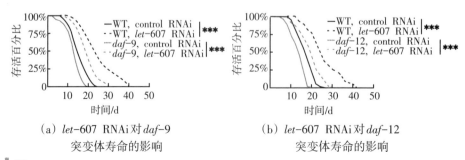

（a）*let*-607 RNAi 对 *daf*-9
　　突变体寿命的影响

（b）*let*-607 RNAi 对 *daf*-12
　　突变体寿命的影响

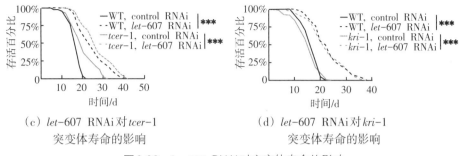

（c）let-607 RNAi对tcer-1
突变体寿命的影响

（d）let-607 RNAi对kri-1
突变体寿命的影响

图3.29 let-607 RNAi对突变体寿命的影响

3.2.5 已知的CREBH调控因子不参与LET-607对DAF-16的调控

哺乳动物CREBH可能受生物钟蛋白（brain and muscle arnt-like 1，BMAL1）和糖原合酶激酶3β（glycogen synthase kinase-3β，GSK3β）的调节[300]。但是，著者发现其线虫直系同源基因aha-1（见图3.30）和gsk-3（见图3.32）的RNAi对MTL-1::GFP表达没有影响（见图3.31和图3.33）。这可能是因为LET-607为核定位蛋白，而BMAL1和GSK3β调节了从内质网到高尔基体的转运，以及随后的CREBH的酶切反应。

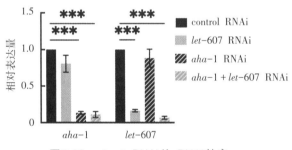

图3.30 aha-1 RNAi的 RNAi效率

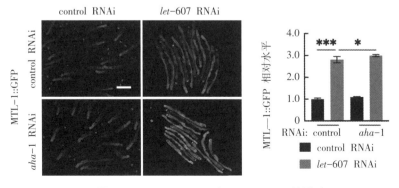

图3.31 aha-1 RNAi对MTL-1::GFP的影响

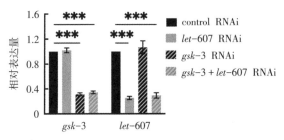

图 3.32 *gsk-3* RNAi 的 RNAi 效率

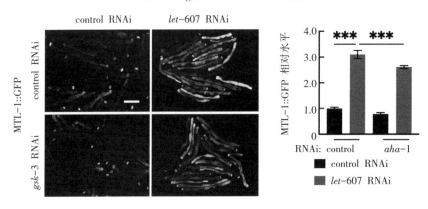

图 3.33 *aha-1* RNAi 对 MTL-1::GFP 的影响

此外，已发现 PPARα 与 CREBH 相互作用以调节靶基因[301]。在此使用 *pparα* 的功能性直系同源基因 *nhr-49* 的突变体线虫与 MTL-1::GFP 线虫进行杂交，得到 *nhr-49*, MTL-1::GFP 线虫进行 *let-607* RNAi 实验。著者发现仅在对照组和 *let-607* RNAi 处理的动物中略微诱导了 MTL-1::GFP 表达（见图 3.34）。

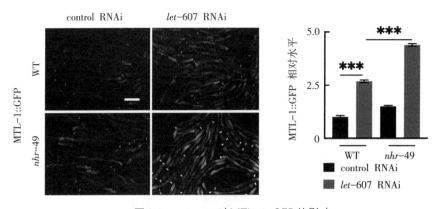

图 3.34 *nhr-49* 对 MTL-1::GFP 的影响

进一步使用突变体线虫 *nhr-49*（*nr2041*）进行活性氧应激实验，发现其与野生型对照没有差别（见图3.35），这些结果表明NHR-49不参与LET-607介导的DAF-16调控。

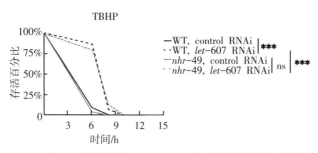

图3.35 *let*-607 RNAi对野生型线虫和*nhr*-49突变体线虫抵抗活性氧应激的影响

本节实验结果显示，LET-607 RNAi通过诱导DAF-16入核将其激活，从而增强秀丽隐杆线虫应激抗性和延长寿命。虽然LET-607参与内质网应激，但是其激活DAF-16却独立于内质网应激系统。而能够与人类CREBH相互作用的BMAL1，GSK3β和PPARα线虫直系同源物AHA-1，GSK-3和NHR-49都不参与LET-607介导的DAF-16的激活体系，这揭示了LET-607与CREBH的不同。*daf*-2和*glp*-1两个依赖*daf*-16的长寿突变体在进行*let*-607 RNAi时呈现不同的结果，本节揭示*let*-607基因RNAi不是通过激活生殖细胞缺失突变体途径来促进长寿的。

3.3 LET-607调控线虫磷脂酰胆碱代谢

3.3.1 LET-607对线虫脂肪累积的影响

let-607 RNAi后的线虫，后代数量、L4时期线虫的体长（见图3.11）都明显少于对照组。而作为储能物质的甘油三酯对线虫的生长发育以及后代繁衍都起着重要作用，所以对其甘油三酯变化进行检测。因为线虫产卵会消耗大量脂肪，本研究针对产卵前的L4时期幼虫进行脂肪染色，*let*-607 RNAi后的线虫较对照组脂肪染色明显变浅（见图3.36~图3.37）。并且将油红O染料萃取后定量结果显示，对照组（control RNAi）线虫含量为18.31 ng/μL，*let*-607 RNAi线虫含量为9.96 ng/μL，差异极显著（$P<0.001$）（见图3.37）。

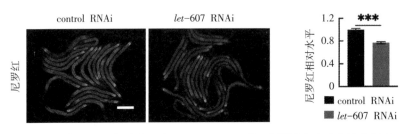

图 3.36　使用尼罗红染色观察 *let*-607 RNAi 对野生型线虫脂肪含量的影响

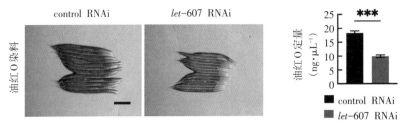

图 3.37　使用油红 O 染料染色观察 *let*-607 RNAi 对野生型线虫脂肪含量的影响

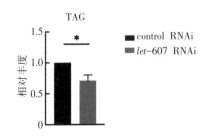

图 3.38　*let*-607 RNAi 对甘油三酯丰度的影响

接下来通过薄层色谱法对线虫的分离，并通过 GC-MS/MS 进行分析。分析结果显示，*let*-607 RNAi 显著降低了线虫甘油三酯的含量（见图 3.38）。

以上实验结果表明，*let*-607 RNAi 后会对线虫的生长发育、后代繁衍产生一定的影响，也会对线虫脂肪的积累产生较大的影响。LET-607 对线虫维持正常生长、发育、繁殖以及脂肪的积累都起着重要作用。

3.3.2　LET-607 对线虫脂肪酸的影响

由于 *let*-607 RNAi 会明显减少线虫的脂肪积累，接下来使用 GC-MS/MS 技术检测野生型线虫体内各游离脂肪酸含量的变化。在对 20000 只 L1 时期野生型线虫进行 *let*-607 RNAi 后，于 L4 时期收集，经过甲酯化处理后用 GC-MS/MS 进行检测，实验结果显示，*let*-607 RNAi 降低了许多脂肪酸的含量（见图 3.39），这导致了饱和脂肪酸和不饱和脂肪酸整体的减少（见图 3.40）。

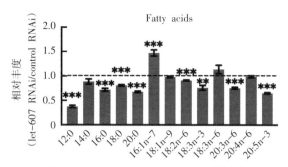

图 3.39 *let*-607 RNAi 对脂肪酸的影响

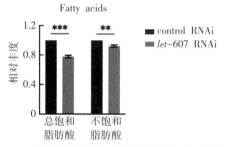

图 3.40 *let*-607 RNAi 对总饱和脂肪酸和不饱和脂肪酸丰度的影响

进一步在 RNA-seq 数据中挖掘出秀丽隐杆线虫中与脂质合成和脂质代谢相关的 135 个基因，可以看出 *let*-607 RNAi 仅显著上调了 *acs*-2 和 *cpt*-3 两个基因的表达量，下调了 *lbp*-8，*lips*-6，*lips*-14 以及 *acdh*-1 的表达量。值得注意的是，在脂肪酸去饱和代谢中的绝大多数基因都有被下调表达的趋势（见图 3.41）。

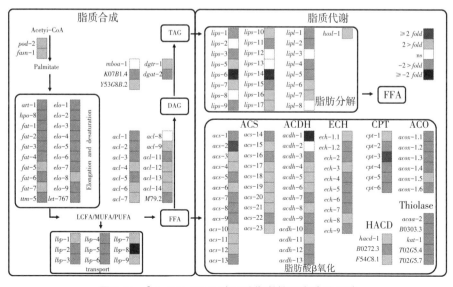

图 3.41 *let*-607 RNAi 对脂质代谢基因表达的影响

结合GC-MS/MS结果中部分脂肪酸的变化，将注意力放到秀丽隐杆线虫多不饱和脂肪酸合成途径（见图3.42）中。使用 fat-1, fat-2, fat-3, fat-4 以及 fat-5 的突变体线虫进行氧化应激和热应激实验。实验结果显示，fat-2 的突变完全抑制了 let-607 基因敲减引起的线虫热应激增强作用；fat-3 和 fat-4 的突变则部分抑制掉这一增强作用；fat-3 的突变可以部分抵消 let-607 基因敲减引起的线虫氧化应激增强作用；fat-2 和 fat-5 的突变则可以完全抑制掉这一增强作用（见图3.43）。

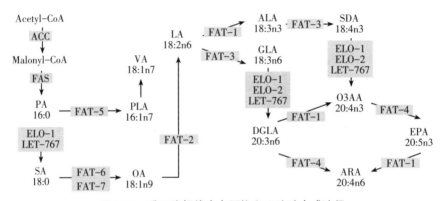

图3.42 秀丽隐杆线虫多不饱和脂肪酸合成途径

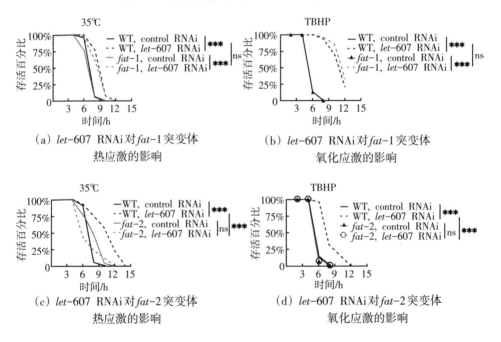

(a) let-607 RNAi对fat-1突变体热应激的影响

(b) let-607 RNAi对fat-1突变体氧化应激的影响

(c) let-607 RNAi对fat-2突变体热应激的影响

(d) let-607 RNAi对fat-2突变体氧化应激的影响

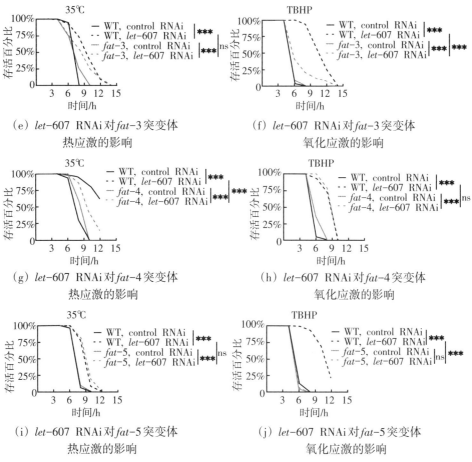

图3.43　let-607 RNAi对fat-1，fat-2，fat-3，fat-4，fat-5突变体热应激和氧化应激的影响

根据以上结果，在只考虑秀丽隐杆线虫多不饱和脂肪酸合成途径单一变量的情况下，推测得知：①假设let-607 RNAi下调了fat-2的表达，理论上造成OA（18:1n9）的积累，LA（18:2n6）减少；从GC-MS/MS结果可以看出，OA（18:1n9）没有变化，LA（18:2n6）减少；同时fat-2的突变体也显示了热应激和氧化应激都被抑制。②假设let-607 RNAi下调了fat-3的表达，理论上造成LA（18:2n6）和ALA（18:3n3）的积累，GLA（18:3n6）和SDA（18:4n3）的减少；从GC-MS/MS结果可以看出，LA（18:2n6）和ALA（18:3n3）都和理论值正相反，是减少的，同时GLA（18:3n6）却也有增加的趋势；但是fat-3的突变体也显示了热应激和氧化应激的部分抑制。③假设let-607 RNAi下调了fat-4的表达，理论上造成O3AA（20:4n3）和DGLA（20:3n6）的积累，EPA（20:5n3）和AA（20:4n6）的减少；从GC-MS/MS结果可以看出，DGLA（20:

3n6)和理论值正相反,是减少的,不过EPA(20:5n3)与推测一样是减少的,并且 *fat-4* 的突变体也显示了热应激的部分抑制。④同时从RNA-seq数据也可以看出,*let-607* RNAi对 *fat-5* 是有上调趋势的,理论上 *fat-5* 上调会造成PA(16:0)的减少和PLA(16:1n7)的增加;这一推测和GC-MS/MS结果是相一致的;但是,*fat-5* 的突变体,理论上会造成PA(16:0)的增加和PLA(16:1n7)的减少,却也变现成对氧化应激的完全抑制作用。⑤*fat-1* 的下调,理论上也会引起LA(18:2n6),DGLA(20:3n6)和AA(20:4n6)的增加,ALA(18:3n3),O3AA(20:4n3)和EPA(20:5n3)的减少;但是这一推测和GC-MS/MS结果是不相同的,并且也与此前的推测相违背。

至此,根据以上结果和推测可以得出一个结论:*let-607* RNAi对线虫的脂肪积累产生了影响,并且有可能通过影响其中某些种类的脂质,而不是某个单一脂肪酸来调控DAF-16的活性,从而使线虫的抗逆性增加和寿命延长。

3.3.3 酸性鞘磷脂酶ASM-3不介导DAF-16激活

在RNA-seq的数据中,*sms-5* 和 *asm-3* 这两个基因引起了著者的注意。它们在线虫神经酰胺代谢的可逆反应中,神经酰胺和磷脂酰胆碱在 *sms-5* 的作用下生成鞘磷脂和二酰基甘油;相反地,鞘磷脂和二酰基甘油又在 *asm-3* 的作用下生成神经酰胺和磷脂酰胆碱(见图3.44)。并且,*let-607* RNAi后 *sms-5* 被显著上调(上调2.29倍),而 *asm-3* 被显著下调(下调9.24倍)。接下来首先对MTL-1::GFP进行 *asm-3* RNAi来验证这一推测。

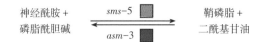

图3.44 SMS-5和ASM-3介导的酶促反应示意图

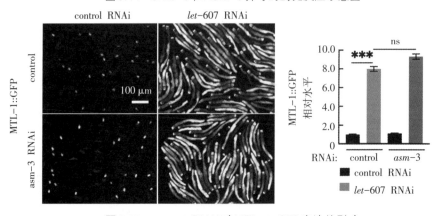

图3.45 *asm-3* RNAi对MTL-1::GFP表达的影响

实验结果显示，asm-3 RNAi 并不会对 MTL-1::GFP 的表达造成任何影响（见图 3.45）。考虑到酸性鞘磷脂酶家族（Acid sphingomyelinase）有 asm-1，asm-2 和 asm-3 三个基因，且都参与鞘磷脂合成神经酰胺的反应，在不确定 asm-1，asm-2 是否因为敲减 asm-3 而起到代偿作用的情况下，使用了酸性鞘磷脂酶的特异性抑制剂地昔帕明（desipramine）进行验证（见图 3.46）。进一步对线虫 asm-3 RNAi 以及施用地昔帕明后进行热应激和氧化应激实验，线虫对应外界压力的抵抗能力也没有任何改善（见图 3.47）。实验结果表明，asm-3 的确不参与 ECSPC。

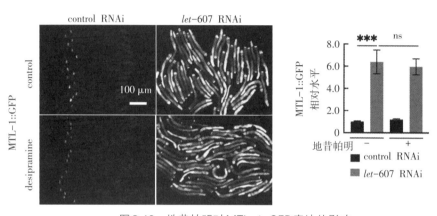

图 3.46 地昔帕明对 MTL-1::GFP 表达的影响

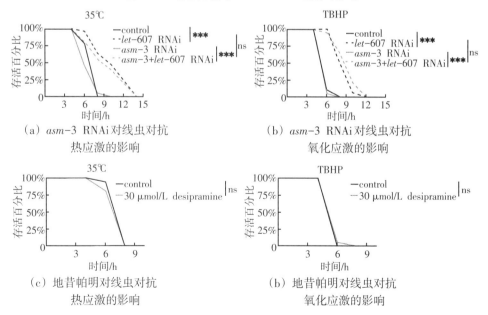

（a）asm-3 RNAi 对线虫对抗
热应激的影响

（b）asm-3 RNAi 对线虫对抗
氧化应激的影响

（c）地昔帕明对线虫对抗
热应激的影响

（b）地昔帕明对线虫对抗
氧化应激的影响

图 3.47 asm-3 RNAi 和地昔帕明对线虫对抗热应激和氧化应激的影响

3.3.4 鞘磷脂合酶SMS-5介导DAF-16激活

通过对MTL-1::GFP进行RNAi，筛选了RNA-seq中全部显著上调表达的基因，结果显示 sms-5 RNAi 明显抑制了在 let-607 RNAi 后 MTL-1::GFP 的激活（见图3.48）。同时，对 let-607 RNAi 的 DAF-16::GFP 线虫同时进行 sms-5 RNAi，发现由 let-607 RNAi 引起的 DAF-16 入核也被抑制了（见图3.49）。

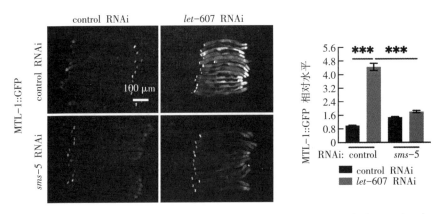

图3.48 sms-5 RNAi 对 let-607 RNAi 诱导的 MTL-1::GFP 表达的影响

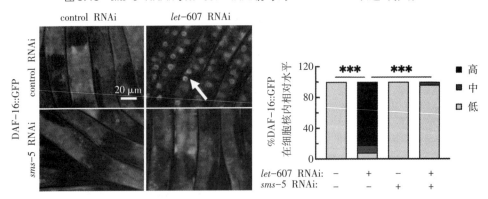

图3.49 sms-5 RNAi 对 let-607 RNAi 诱导的 DAF-16::GFP 入核的影响

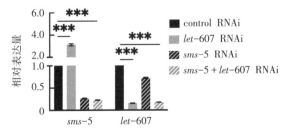

图3.50 let-607 RNAi 和 sms-5 RNAi 对 sms-5 mRNA 表达量的影响

为此，我们用野生型线虫进行 let-607 RNAi 后提取的 RNA 进行了 QPCR 来验证 sms-5 的表达量是否与 RNA-seq 结果一致。如图 3.44 所示，sms-5 明显上调表达。同时对野生型线虫进行 sms-5 RNAi 处理后提取 RNA 验证 sms-5 的表达量，以此检测 sms-5 的 RNAi 效率。结果显示，sms-5 的 RNAi 效率可达约 78%（见图 3.50），符合实验需求。

下一步验证 sms-5 RNAi 后线虫的抗氧化应激能力和寿命，sms-5 RNAi 后的线虫也表现出了部分抑制 let-607 RNAi 增强的线虫抗氧化应激能力（见图 3.51）和寿命（见图 3.52）。

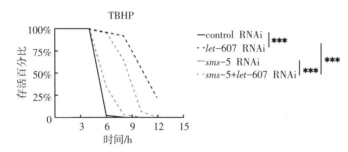

图 3.51 sms-5 RNAi 对 let-607 RNAi 诱导的抗氧化应激的影响

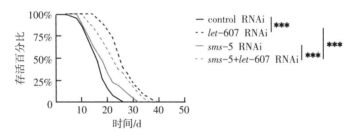

图 3.52 sms-5 RNAi 对 let-607 RNAi 诱导的寿命的影响

上述实验结果可以推导出，秀丽隐杆线虫鞘磷脂合酶 SMS-5 是 DAF-16 的新调节子，是 LET-607 激活 DAF-16 所必需的。

3.3.5 磷脂酰胆碱介导 DAF-16 激活

由于在秀丽隐杆线虫细胞中，鞘磷脂合成酶 SMS-5 以神经酰胺和磷脂酰胆碱为底物，从而生成鞘磷脂和二酰基甘油（见图 3.53），因此可以推测这些膜脂质的变化是 DAF-16 活化的原因。

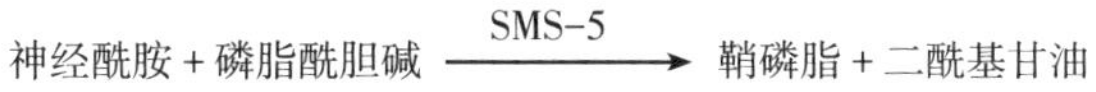

$$\text{神经酰胺} + \text{磷脂酰胆碱} \xrightarrow{\text{SMS-5}} \text{鞘磷脂} + \text{二酰基甘油}$$

图 3.53 SMS-5 介导的酶促反应示意图

研究结果表明,膜脂质特别是鞘脂,是人类衰老的生物标志物,长寿人群的血清分析显示特定鞘脂的增加[260]。而与野生型小鼠相比,长寿裸鼠中的几种血浆鞘脂及其代谢产物也增加了[302]。鞘脂可以转化为鞘氨醇-1-磷酸(sphingosine-1-phosphate,S1P)和神经酰胺两种具有相反生物学作用的生物活性脂质,S1P促进细胞增殖和存活,而神经酰胺则促进细胞凋亡[303-305]。S1P和神经酰胺之间的平衡称为S1P/神经酰胺轴,可调节衰老过程[305]。神经酰胺会在衰老的线虫和人类中积聚[306]。此外,富含神经酰胺的饮食会缩短秀丽隐杆线虫的寿命[307]。相反,缺乏合成神经酰胺酶的线虫突变体是长寿的[308]。此前已有报道特定种类的因为神经酰胺在线虫抗缺氧性[309]和寿命延长[310]方面有特定的作用,并且神经酰胺也参与调节应激诱导的细胞凋亡[311]。由此推测,是否 *let*-607 RNAi导致了神经酰胺的减少,从而促使一系列增强反应。因此,著者首先直接通过回补神经酰胺的方式来验证是否是神经酰胺的减少导致了 *let*-607介导的DAF-16激活。结果表明,在有或者没有 *let*-607 RNAi的情

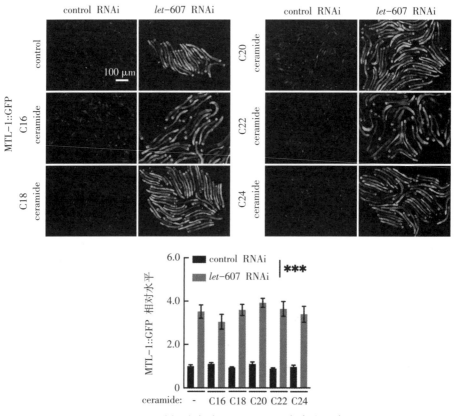

图3.54 神经酰胺对MTL-1::GFP表达的影响

况下，各类型的神经酰胺对MTL-1::GFP的表达都没有造成影响（见图3.54），说明神经酰胺不参与 let-607 介导的DAF-16激活。

接下来验证是否由于磷脂酰胆碱的减少而激活了DAF-16。通过在线虫食物中补充胆碱（一种磷脂酰胆碱合成前体），可以观察到胆碱可以抑制在 let-607 RNAi后MTL-1::GFP的表达（见图3.55）。

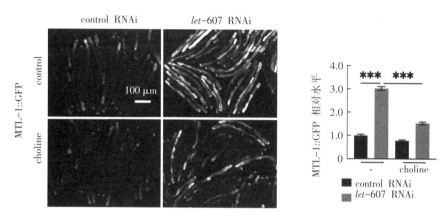

图3.55　胆碱对 let-607 RNAi诱导的MTL-1::GFP表达的影响

在秀丽隐杆线虫中，磷脂酰胆碱是胆碱通过肯尼迪途径（Kennedy pathway）合成的，或者是由SAMS-1和PMT-1/PMT-2介导的磷酸乙醇胺的甲基化生成的（见图3.56）。

图3.56　秀丽隐杆线虫磷脂酰胆碱生物合成途径

接下来，验证减少磷脂酰胆碱含量是否足以通过RNAi磷脂酰胆碱合成所需的基因来激活DAF-16。结果显示，磷脂酰胆碱生物合成基因（包括 sams-1，ckb-1，pmt-2，pcyt-1 和 cept-1）的RNAi对MTL-1::GFP表达没有影响或影响不大（见图3.57）。此外，测试了 let-607 和磷脂酰胆碱生物合成基因的协同作用，发现 let-607 和每个磷脂酰胆碱合成基因通过双重组合RNAi可以协同作用叠加激活MTL-1::GFP（见图3.57），推测减少磷脂酰胆碱可以介导DAF-16激活。

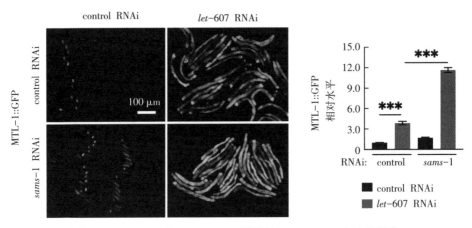

（a）sams-1 RNAi对let-607 RNAi诱导的MTL-1::GFP表达的影响

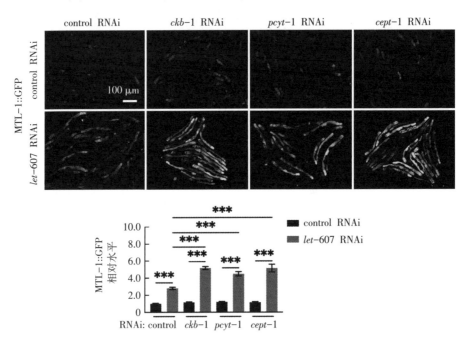

（b）ckb-1 RNAi, pcyt-1 RNAi和cept-1 RNAi对let-607 RNAi诱导的MTL-1::GFP表达的影响

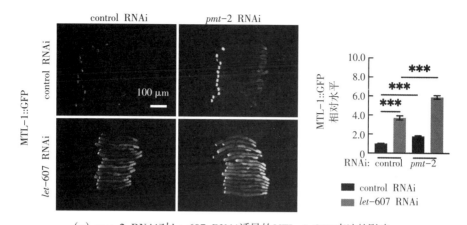

(c) *pmt-2* RNAi对*let-607* RNAi诱导的MTL-1::GFP表达的影响

图3.57 磷脂酰胆碱合成基因RNAi对*let-607* RNAi诱导的MTL-1::GFP表达的影响

同时，针对double RNAi系统也进行了QPCR，验证了RNAi系统的RNAi效率（见图3.58），结果显示各基因RNAi效率都有效。

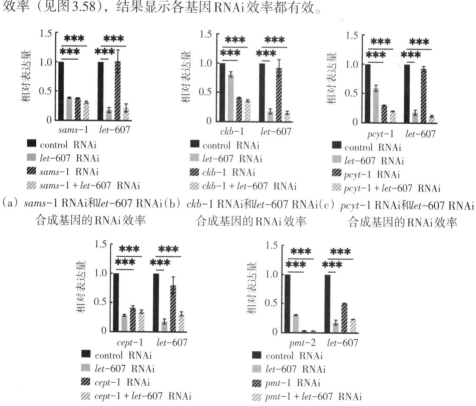

(a) *sams-1* RNAi和*let-607* RNAi 合成基因的RNAi效率
(b) *ckb-1* RNAi和*let-607* RNAi 合成基因的RNAi效率
(c) *pcyt-1* RNAi和*let-607* RNAi 合成基因的RNAi效率
(d) *cept-1* RNAi和*let-607* RNAi 合成基因的RNAi效率
(e) *pcyt-2* RNAi和*let-607* RNAi 合成基因的RNAi效率

图3.58 PC生物合成基因RNAi的 RNAi效率

综上所述,这些数据与SMS-5介导的磷脂酰胆碱减少对于响应 let-607 RNAi 的 DAF-16 激活是至关重要的。

3.3.6 LET-607调控磷脂酰胆碱代谢

此前对线虫游离脂肪酸进行的GC-MS/MS结果已经得知,let-607 RNAi 导致许多脂肪酸种类的含量降低(见图3.39),从而导致饱和脂肪酸和不饱和脂肪酸整体含量下降(见图3.40)。接下来通过薄层色谱法对线虫四种磷脂(phospholipids,PLs)进行分离(磷脂酰胆碱,PC;磷脂酰乙醇胺,PE;磷脂酰肌醇,PI;磷脂酰丝氨酸,PS),并通过GC-MS/MS进行分析。分析结果显示,let-607 RNAi 显著降低了磷脂酰胆碱的含量(见图3.59),并且主要是由于不饱和磷脂酰胆碱的减少(见图3.60)。

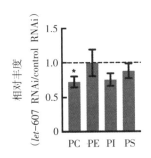

图3.59 *let*-607 RNAi对膜磷脂丰度的影响

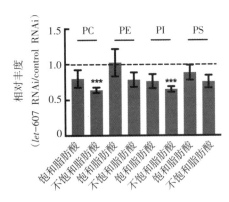

图3.60 *let*-607 RNAi对PC,PE,PI,PS饱和与不饱和脂肪酸丰度的影响

此外,检测了其他三种主要的膜磷脂,结果表明磷脂酰乙醇胺和磷脂酰丝氨酸不受显著影响,而磷脂酰肌醇,特别是不饱和磷脂酰肌醇降低了(见图3.60)。还计算了可能影响膜特性(包括流动性和曲率)的PC/PE比值,发现

它在 let-607 RNAi 线虫中显示出下降的趋势（见图3.61）。

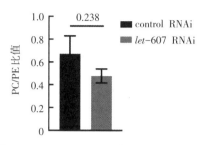

图3.61 let-607 RNAi 对 PC/PE 比值的影响

进一步详细分析了磷脂酰胆碱的脂肪酸组成，发现其中的许多不饱和脂肪酸确实大大减少（见图3.62），作为对照，磷脂酰乙醇胺中的大多数不饱和脂肪酸均无明显变化（见图3.63）。这一结果再次证明了 LET-607 在 PC 调节中的重要性。

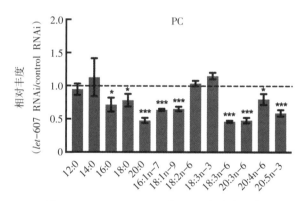

图3.62 let-607 RNAi 对 PC 脂肪酸的影响

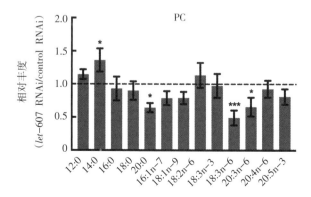

图3.63 let-607 RNAi 对 PE 脂肪酸的影响

接下来我们直接回补各种类磷脂酰胆碱,结果显示只有PC(18:1n9)可以抑制由 let-607 RNAi 诱导的 MTL-1::GFP(见图3.64),而PC(18:1n9)的结果也复刻了 sms-5 RNAi 的结果,并暗示不饱和PC可以调节DAF-16。

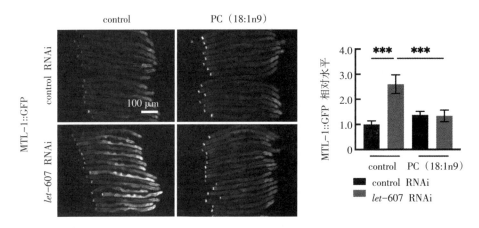

图3.64 回补PC(18:1n9)对 let-607 RNAi 诱导的 MTL-1::GFP 表达的影响

同时,对N2进行了相同的PC(18:1n9)回补实验,再进行TLC来验证线虫是否对回补的PC(18:1n9)有吸收作用,结果显示回补的PC(18:1n9)确实被线虫吸收了(见图3.65)。

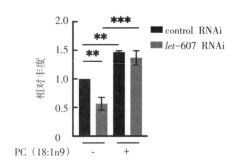

图3.65 回补PC(18:1n9)对线虫磷脂酰胆碱中18:1n9脂肪酸的影响

综合以上结果,可以推断PC的减少确实可以介导DAF-16的激活,并且由SMS-5介导的特定类型的磷脂酰胆碱减少对于响应 let-607 RNAi 对DAF-16的激活是至关重要的。

3.3.7 PC介导生殖细胞缺失型突变体中的DAF-16激活

在建立了SMS-5和PC作为DAF-16激活介体的模型后,接着验证在其他模型中DAF-16的激活是否需要它们,包括 *daf-2*（*e*1368）突变体、*glp-1*（*e*2141）突变体、热应激和氧化应激。

有趣的是,不饱和PC（18:1n9）会特异地在 *glp-1* 突变体（见图3.66）中抑制MTL-1::GFP的诱导,而在 *daf-2* 突变体（见图3.67）以及遭受热应激（见图3.68）和遭受氧化应激（见图3.69）的线虫中则没有抑制MTL-1::GFP的诱导。

这些数据表明PC代谢可能参与控制 *glp-1* 突变体中DAF-16的活化。

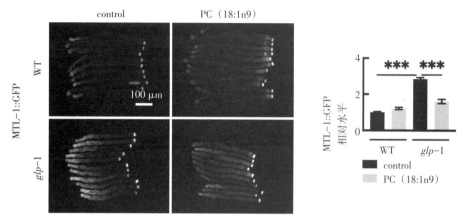

图3.66　回补PC（18:1n9）对 *glp-1* 突变体中MTL-1::GFP诱导的影响

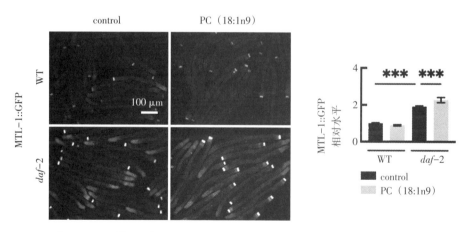

图3.67　回补PC（18:1n9）对 *daf-2* 突变体中MTL-1::GFP诱导的影响

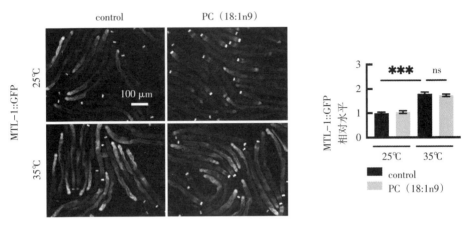

图3.68　回补PC（18:1n9）对35℃热应激后野生型线虫中MTL-1::GFP诱导的影响

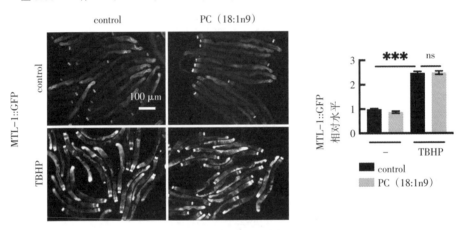

图3.69　回补PC（18:1n9）对TBHP处理后野生型线虫中MTL-1::GFP诱导的影响

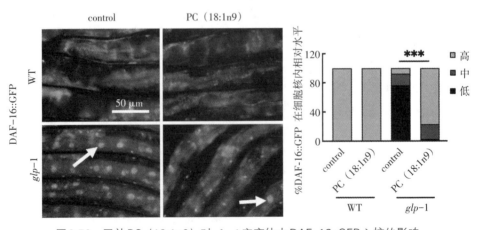

图3.70　回补PC（18:1n9）对 glp-1 突变体中DAF-16::GFP入核的影响

同时观察各条件下DAF-16::GFP的入核情况，这也显示出与此前相似的结果（见图3.70和图3.71），表明PC特异性介导生殖细胞缺失突变体中DAF-16的活化，与寿命数据一致［见图3.29（b）］。

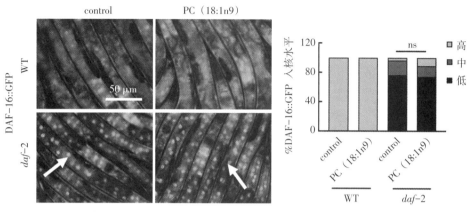

图3.71　回补PC（18:1n9）对 daf-2 突变体中DAF-16::GFP入核的影响

至于 sms-5，我们发现它的RNAi在 glp-1 突变体（见图3.72）和 daf-2 突变体（见图3.73），以及热刺激（见图3.74）和活性氧刺激的压力条件下（见图3.75）均不影响MTL-1::GFP的诱导。

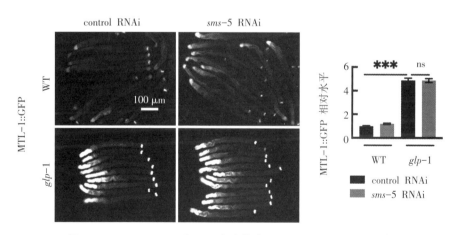

图3.72　sms-5 RNAi对 glp-1 突变体中MTL-1::GFP诱导的影响

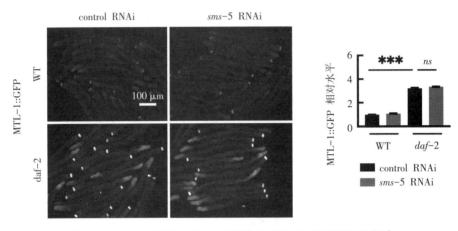

图3.73 sms-5 RNAi对daf-2突变体中MTL-1::GFP诱导的影响

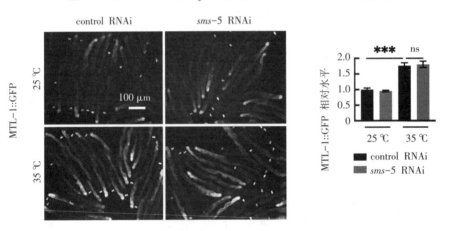

图3.74 sms-5 RNAi对35℃热刺激的野生型线虫中MTL-1::GFP诱导的影响

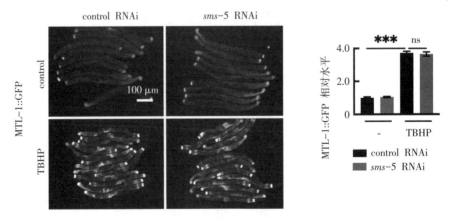

图3.75 sms-5 RNAi对TBHP处理的野生型线虫中MTL-1::GFP诱导的影响

sms-5 RNAi可以抑制*let*-607 RNAi后激活的DAF-16，从而削减其增强应激抗性和延长寿命的作用。*let*-607 RNAi会减少线虫肠道内储存的脂质，并且对其饱和脂肪酸和不饱和脂肪酸也有减少作用。而SMS-5通过催化磷脂酰胆碱代谢而使线虫磷脂酰胆碱减少，本节通过TLC实验检测到*let*-607 RNAi后线虫的磷脂酰胆碱，特别是不饱和磷脂酰胆碱急剧减少，并且通过回补PC（18：1n9）可以抑制*let*-607 RNAi对MTL-1::GFP的激活作用，从而确定了*sms*-5和磷脂酰胆碱是LET-607激活DAF-16中的关键中间介质，并且进一步发现磷脂酰搭建代谢可能参与控制*glp*-1突变体中DAF-16的活化。

3.4　LET-607通过钙离子信号调控DAF-16

作为最丰富的膜脂质之一，磷脂酰胆碱对膜蛋白有重要影响。磷脂酰胆碱的含量和组成会影响膜特性，进而改变膜蛋白的功能，例如磷脂酰胆碱的变化导致内质网膜定位蛋白活性的变化[312-313]。内质网是蛋白质和脂质合成以及细胞内钙存储的主要部位，改变内质网稳态会激活未折叠蛋白反应[8]。定位在内质网上的钙离子通道和钙稳态可通过磷脂酰胆碱含量的变化来调节[313]，据报道钙信号参与DAF-16活性的调节[185]。因此，我们下一步验证LET-607激活DAF-16是否需要内质网膜上锚定的钙离子通道参与。

3.4.1　ITR-1钙离子通道对DAF-16激活的影响

肌醇1,4,5-三磷酸（inositol 1,4,5-trisphosphate，IP3）受体将钙离子从内质网释放到细胞质中，而肌浆网/内质网Ca^{2+}-ATPase（sarco/endoplasmic reticulum Ca^{2+}-ATPase，SERCA）则将钙离子从细胞质泵入内质网内。在秀丽隐杆线虫中，IP3和SERCA受体分别由*itr*-1和*sca*-1编码。

首先验证ITR-1钙离子通道，推测*itr*-1的缺失会减少细胞质内钙离子浓度，从而影响DAF-16的激活。观察到*itr*-1 RNAi在*let*-607 RNAi后基本抑制了MTL-1::GFP的激活（见图3.76）。此外，*itr*-1缺失突变体可显著消除*let*-607 RNAi线虫的抗逆性［见图3.77（a）］和寿命［见图3.77（b）］，同时也检测了*itr*-1的RNAi效率（见图3.78）。

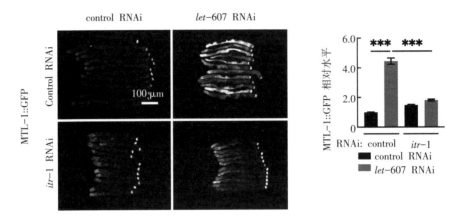

图3.76 *itr-1* RNAi对*let-607* RNAi诱导的MTL-1::GFP表达的影响

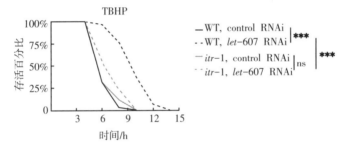

（a）*itr-1*突变体对*let-607* RNAi诱导的活性氧抗性的影响

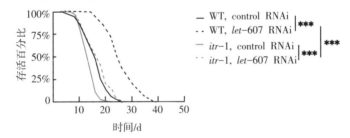

（b）*itr-1*突变体对*let-607* RNAi诱导的寿命的影响

图3.77 *itr-1*突变体对*let-607* RNAi诱导的活性氧抗性和寿命的影响

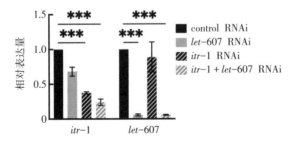

图3.78 *let-607* RNAi和*itr-1* RNAi对*itr-1* mRNA表达量的影响

此前有研究表明，EGL-8参与了真菌感染诱导的EGL-8生成IP3，激活内质网膜上IP3受体ITR-1钙离子通道，释放钙离子到细胞质后经过一系列反应，最终激活DAF-16从而增强线虫对真菌感染的先天免疫[314]。在秀丽隐杆线虫中EGL-8编码磷脂酶C β（phospholipase C β），参与将磷脂酰肌醇4,5-双磷酸酯（PIP2）裂解为小的信号分子肌醇1,4,5-三磷酸酯（IP3）和二酰基甘油（DAG）。接下来先验证通过 *egl*-8 RNAi验证降低ITR-1的配体IP3，从而降低的ITR-1依赖性钙信号传导是否参与到LET-607介导的DAF-16激活。验证结果表明，*egl*-8 RNAi完全消除了 *let*-607 RNAi对MTL-1::GFP表达（见图3.79）和氧化应激抗性（见图3.80）的影响，这与 *itr*-1突变体的结果一致，同时我们也检测了 *egl*-8的RNAi效率（见图3.81）。

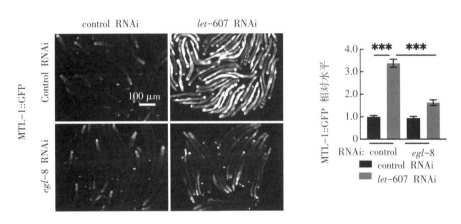

图3.79 *egl*-8 RNAi对 *let*-607 RNAi诱导的MTL-1::GFP表达的影响

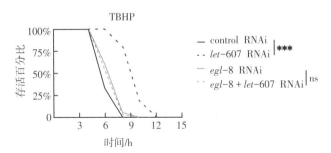

图3.80 *egl*-8 RNAi对 *let*-607 RNAi诱导的活性氧抗性的影响

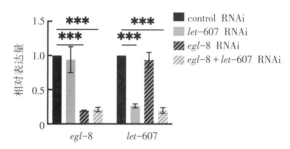

图3.81 *let*-607 RNAi和*egl*-8 RNAi对*egl*-8 mRNA表达量的影响

egl-8 RNAi可以减少IP3和DAG，这一过程抑制了LET-607介导的DAF-16的激活，而*let*-607敲减后上调的*sms*-5通过消耗PC生成DAG则可以激活DAF-16。随后通过敲减秀丽隐杆线虫二酰基甘油激酶家族5个成员*dgk*-1，*dgk*-2，*dgk*-3，*dgk*-4，*dgk*-5来造成线虫体内DAG的积累，从而观察DAG对MTL-1::GFP的影响（见图3.82）。

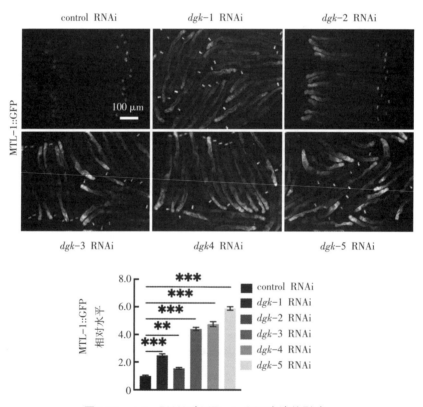

图3.82 *dgk*s RNAi对MTL-1::GFP表达的影响

同时，我们使用1-油酰基-2-乙酰基-sn-甘油（一种二酰基甘油）来提高细胞中的二酰基甘油,从而激活 MTL-1::GFP 的表达(见图 3.83)。可以发现,无论是敲减 *dgk*s 还是回补 DAG,都能使 MTL-1::GFP 被激活。

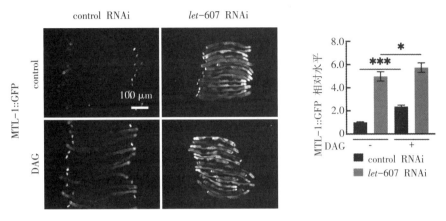

图 3.83　DAG 对 MTL-1::GFP 表达的影响

这些数据表明 LET-607 介导的 DAF-16 激活需要 ITR-1 钙离子通道的参与。

3.4.2　钙离子流对 DAF-16 激活的影响

随后著者检验了 SERCA,发现 *sca*-1 RNAi 增加了 MTL-1::GFP 的强度(见图 3.84),更重要的是,*sca*-1 RNAi 可以在 *sms*-5 RNAi 的基础上恢复 MTL-1::GFP 的表达(见图 3.84),这表示钙信号作用于 SMS-5 的下游。同时,也用 QPCR 验证了各 RNAi 的效率,以确保实验真实有效(见图 3.85)。

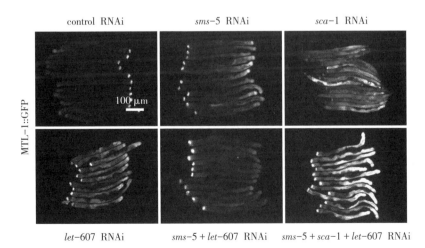

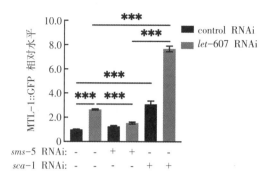

图3.84 *sca-1* RNAi对*sms-5* RNAi抑制MTL-1::GFP表达的影响

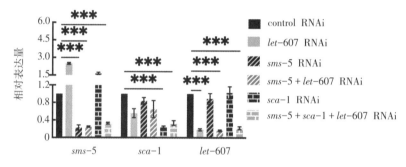

图3.85 图3.84中各基因RNAi效率

SCA-1负责将细胞质中的钙离子运输到内质网中，由此推测*sca-1* RNAi是否导致细胞质内钙离子浓度降低而产生这一现象，CRT-1是作为钙网蛋白存在于内质网内而行使贮存钙离子的作用的，随即对MTL-1::GFP线虫*crt-1*进行RNAi使钙离子从内质网释放到细胞质来模拟这一推测，发现*crt-1*在RNAi后也会激活MTL-1::GFP的表达（见图3.86）。

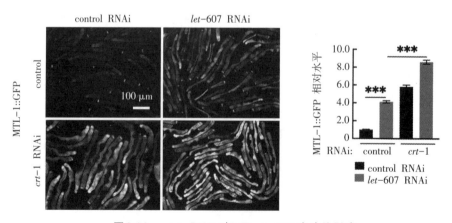

图3.86 *crt-1* RNAi对MTL-1::GFP表达的影响

并且，通过添加离子霉素（一种可以诱导细胞内钙通量的钙离子载体）模拟 sca-1 RNAi，发现离子霉素的刺激可响应 sms-5 RNAi 逆转 MTL-1::GFP 的表达（见图3.87），这一结果和 sca-1 RNAi 结果一致。

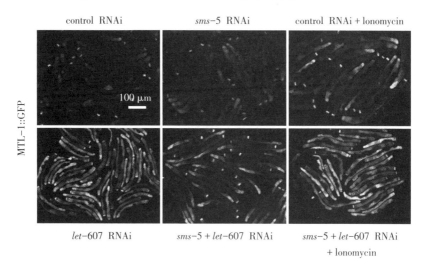

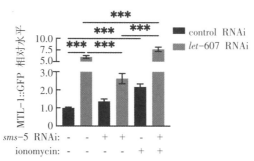

图3.87　离子霉素对被 sms-5 RNAi 抑制的 MTL-1::GFP 表达的影响

3.4.3　PKC-2和SGK-1参与LET-607介导的DAF-16激活

此前也有报道，线虫通过钙离子激活PKC-2及其下游激酶SGK-1来调节DAF-16活性[185]。根据前人研究，著者使用编码蛋白激酶Cα亚基的 pkc-2 来进行验证[314]。观察到 pkc-2 RNAi 可以完全抑制 let-607 RNAi 激活的 MTL-1::GFP（见图3.88）。同时，用QPCR验证了 pkc-2 RNAi 的效率，以确保实验真实有效（见图3.89）。

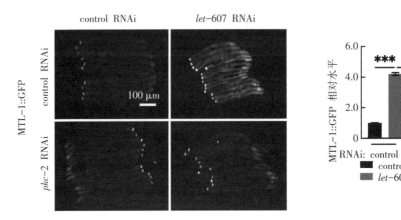

图3.88 *pkc-2* RNAi对*let-607* RNAi诱导的MTL-1::GFP表达的影响

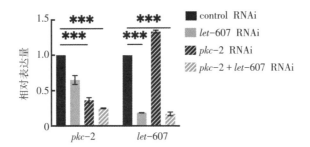

图3.89 *let-607* RNAi和*pkc-2* RNAi对*pkc-2* mRNA表达量的影响

接着观察到*sgk-1* RNAi可以完全抑制*let-607* RNAi激活的MTL-1::GFP（见图3.90），同时使用*sgk-1*（*mg*455）突变体线虫与MTL-1::GFP杂交后进行*let-607* RNAi也复刻了*sgk-1* RNAi实验（见图3.91）。

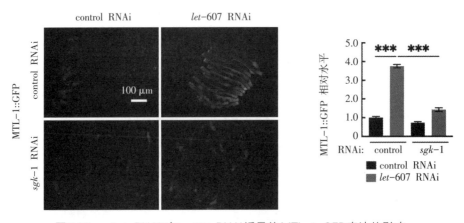

图3.90 *sgk-1* RNAi对*let-607* RNAi诱导的MTL-1::GFP表达的影响

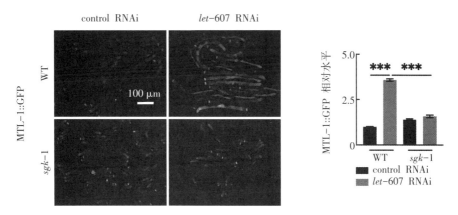

图 3.91 let-607 RNAi 对 sgk-1；MTL-1::GFP 表达的影响

使用 sgk-1（mg455）突变体线虫进行活性氧应激实验发现，sgk-1 的缺失可以抑制 let-607 RNAi 时抗逆性的增强［见图 3.92（a）］；同时，对野生型线虫 N2 进行 sgk-1 RNAi 也可以在 let-607 RNAi 时使抗逆性增强［见图 3.92（b）］，并使寿命延长（见图 3.93）。

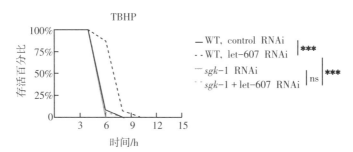

（a）sgk-1 突变体对 let-607 RNAi 诱导的活性氧抗性的影响

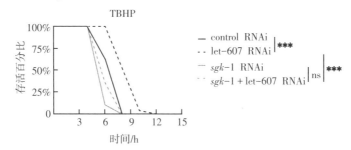

（b）sgk-1 RNAi 对 let-607 RNAi 诱导的活性氧抗性的影响

图 3.92 sgk-1 突变体和 sgk-1 RNAi 对 let-607 RNAi 诱导的活性氧抗性的影响

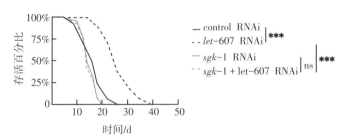

图3.93 *sgk-1* RNAi对*let-607* RNAi诱导的寿命的影响

前期研究表明TORC2组分*rict-*1/Rictor是秀丽隐杆线虫中*sgk-*1的上游调控因子[192-193]。于是著者用*rict-*1（*mg*451）突变体线虫和MTL-1::GFP进行杂交，并进行*let-*607 RNAi处理。但是，与*sgk-*1不同，*rict-*1的突变并未抑制*let-*607 RNAi处理的蠕虫中的MTL-1::GFP诱导（见图3.94）；同时，用*rict-*1（*mg*451）突变体线虫进行氧化应激实验，也发现*rict-*1的突变同样无作用（见图3.95）。这表明LET-607调节DAF-16的激活不涉及TORC2途径。

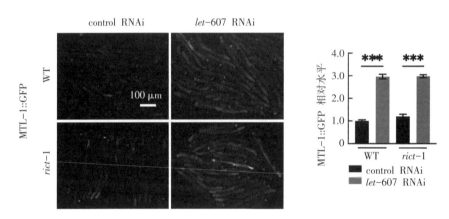

图3.94 *let-*607 RNAi对*rict-*1；MTL-1::GFP表达的影响

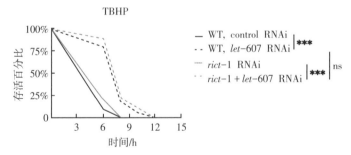

图3.95 *rict-*1突变体对*let-*607 RNAi诱导的活性氧抗性的影响

DAF-16也可以被AKT磷酸化而抑制[135,315]。于是用 akt-1（mg306）突变体线虫和MTL-1::GFP杂交后进行 let-607 RNAi处理，发现 akt-1 和 akt-2（RNAi）增强了MTL-1::GFP的表达，但在野生型和 let-607 RNAi动物中均具有相似的作用（见图3.96），这表明AKT与LET-607介导的DAF-16激活无关。

以上数据表明，let-607 RNAi介导的DAF-16激活需要PKC-2和SGK-1激酶。

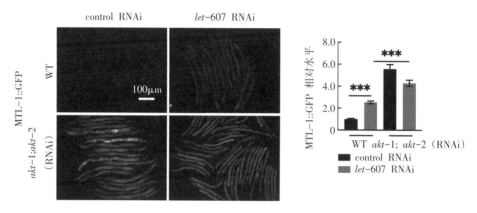

图3.96 let-607 RNAi对 akt-1；akt-2（RNAi）；MTL-1::GFP表达的影响

本章发现内质网上的关键钙离子通道ITR-1参与LET-607介导的DAF-16激活系统。并且通过进一步研究发现 let-607 RNAi介导的DAF-16激活还需要PKC-2和SGK-1，而这一系统又独立于AKT-1激活DAF-16的作用。

4 讨 论

　　膜脂主要由磷脂、鞘脂和胆固醇等组成，是细胞内含量最丰富的脂类分子，同时也是生物膜的主要结构组分。近年来研究发现，膜脂不仅起着膜结构单元的作用，其组成对细胞信号转导、细胞功能，甚至人类疾病都有非常重要的影响[261-262]。近年来对人类和模式动物的研究显示，多种膜脂分子含量的变化伴随着生物体的衰老[263]，膜脂在衰老调控中可能产生的重要作用并因此开始受到关注。研究结果表明，膜脂中含量最丰富的磷脂酰胆碱随着年龄的增长出现了多种磷脂酰胆碱含量的增加[256,260]。然而迄今为止，膜脂在衰老调控中的功能和分子机制在很大程度上还不清楚。另外，鉴于膜脂的细胞生物学功能研究尚处起步阶段，解析膜脂的新功能和机制是当前代谢研究领域的前沿和难点。

　　CREBH是CREB3家族成员之一，受控膜内蛋白酶解激活的内质网转录因子家族[266]，而经典的内质网未折叠蛋白反应的调控因子IRE-1，PEK-1和ATF-6则不能激活DAF-16。本书的研究内容展示了秀丽隐杆线虫CREBH同源蛋白LET-607特异地激活了线虫FOXO同源蛋白DAF-16，并进一步以DAF-16依赖的方式促进线虫氧化应激抵抗能力的增强及寿命的延长，从而揭示了LET-607是新的衰老调控因子。作为转录因子，LET-607可能通过调控目的基因表达来影响衰老进程，利用ChIP-seq数据，对LET-607结合的基因与let-607 RNAi上调的DAF-16靶基因进行比对，发现只有sodh-1这一个基因包含LET-607结合位点，表明LET-607不能直接调控DAF-16的靶基因，其调控功能很可能是通过激活DAF-16起作用的。已有的研究结果表明，哺乳动物CREBH的功能是调节炎症基因和铁调素[267,316]，本书通过RNA-seq的数据揭示，秀丽隐杆线虫LET-607诱导了包括C型凝集素和细胞色素P450在内的几个重要的防御基因家族上调表达，这些发现表明了LET-607/CREBH在功能上具有保守性。

　　细胞内质网在应激状态下可能会促进多重防御机制，这些作用不仅有助于恢复内质网自身的功能，还对整个细胞稳态都起着重要的作用，所以内质网在协调细胞的应激反应中起着核心作用。在内质网应激反应中起着关键作用的是经典的内质网未折叠蛋白反应的调控因子IRE-1，PEK-1和ATF-6，以及近些年发现的CREBH/LET-607[10,266,317]。当这些内质网应激反应因子受到损害时，为了恢复细胞稳态可能在除内质网之外的位置参与调控适应性的应激反应。本书发现线虫LET-607位于细胞核中，尽管有别于哺乳动物CREBH锚定在内质网，但它也会受到内质网应激的调节[316]。因此，LET-607与DAF-16之间的交流可能自适应地响应由LET-607特异介导的内质网应激反应，所以LET-607依赖的内质网应激反应途径的鉴定值得进一步研究。

先前的研究报道表明，CREBH在葡萄糖和脂质代谢中也起着至关重要的作用[236,266]。本书揭示了LET-607在膜脂质代谢中的新功能，因此推测LET-607可能对膜脂质紊乱作出反应并参与膜稳态的维持。利用RNA-seq进行转录组分析，鉴定出受LET-607调控的860个差异表达基因，针对这些差异表达的基因进一步利用RNAi筛选鉴定出介导LET-607调控功能的关键基因为鞘磷脂合酶基因 sms-5 (sphingomyelin synthase 5)，其功能是催化神经酰胺和磷脂酰胆碱代谢为二酰基甘油和鞘磷脂，所有这些膜脂质都是多种细胞过程的关键介质，这暗示着LET-607在代谢和细胞体内平衡中可能发挥更广泛的作用。本书发现 let-607 RNAi可增加 sms-5 基因的表达，而 sms-5 RNAi则能抑制 let-607 RNAi诱导的DAF-16激活、氧化应激抵抗增强和寿命延长。SMS-5作为转移酶参与调控膜脂平衡，通过生化和遗传分析，本书发现在 let-607 RNAi的线虫体内，sms-5 的激活导致的不饱和PC含量下降是激活DAF-16的主要原因，SMS-5和膜脂质磷脂酰胆碱调节DAF-16并诱导长寿。值得注意的是，Lin等通过长期施用D-半乳糖建立了加速大脑衰老的小鼠模型，随后对小鼠海马体进行了多平台代谢组学分析，发现磷脂酰乙醇胺随着小鼠海马体衰老过程而减少；反之，磷脂酰胆碱随着衰老而大量积累[256]。Montoliu等结合核磁共振代谢组学和鸟枪法脂质组学对比分析来自意大利的98名百岁老人［平均年龄（100.7±2.1）岁］和196名普通老人［平均年龄（70±6）岁］的血清，发现与普通老人相比，百岁老人饱和的两种磷脂酰胆碱降低，多种不饱和磷脂酰胆碱增加，致使他们饱和磷脂酰胆碱/不饱和磷脂酰胆碱的比例增加[260]。这与本书观察到在 let-607 RNAi后，磷脂酰胆碱的减少对动物健康有益的观察一致，但磷脂酰胆碱如何调节老化过程尚未完全了解，具体调控机制还需要通过更多深入的研究来发现。

由于LET-607在内质网应激反应中起着关键的作用，而定位在内质网上的钙离子通道和钙稳态可通过磷脂酰胆碱含量的变化来调节[313]，本书通过遗传学手段揭示了DAF-16的激活是由内质网驻留钙通道蛋白ITR-1介导的，推测特定的磷脂酰胆碱可能通过直接影响ITR-1的功能或通过改变一般的膜特性来发挥作用。也有报道揭示钙信号通过激活PKC-2及其下游激酶SGK-1来参与调节DAF-16的活性[185]。同时本书研究结果也发现 let-607 RNAi 介导的DAF-16激活还需要PKC-2和SGK-1激酶，并且这一系统又独立于AKT-1激活DAF-16的作用。而此前已有研究者猜测包括热激和氧化应激在内的其他细胞毒性应激对DAF-16的激活可能减弱了通过破坏内质网稳态而直接激活DAF-16的进化力。DAF-16的这种独立于内质网的激活系统可能为跨压力兴奋奠定了基础，甚至在内质网稳态失衡之前就提高了动物应对内质网应激的能力[137]。这些发现都进一步夯实了LET-607通过调控DAF-16从而促进线虫应激抵抗增强和寿命延长的证据。

本书揭示了应激反应途径之间的新型交流系统，并确定了一个新型的衰老调控因子LET-607，以及SMS-5介导的膜脂质磷脂酰胆碱的调节，该调节通过ITR-1依赖性钙信号传导以及下游激酶PKC-2和SGK-1调节DAF-16途径（见图4.1）。并且通过实验发现，不管是发育前还是发育后对线虫进行 let-607 RNAi，都可以增强其氧化应激能力、热应激能力和寿命。这一研究结果表明LET-607在发育和长寿中起作用，并具有基因多效性。

该发现初步建立了磷脂酰胆碱与寿命的因果关系，为磷脂酰胆碱调控衰老的机制研究提供了一定的理论基础。考虑到本书中发现的蛋白质和脂质的保守性，类似的机制可能在其他物种（如哺乳动物）的细胞稳态和的寿命中起作用。

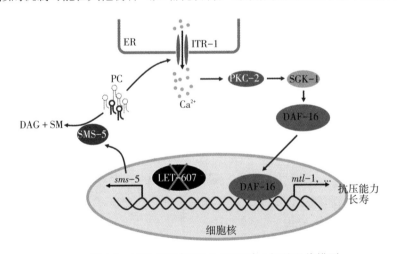

图4.1　LET-607与DAF-16通路之间的通信模型

参考文献

[1] BALCH W E, MORIMOTO R I, DILLIN A, et al. Adapting proteostasis for disease intervention[J]. Science, 2008, 319(5865): 916-919.

[2] HAIGIS M C, YANKNER B A. The aging stress response[J]. Molecular cell, 2010, 40(2): 333-344.

[3] KOURTIS N, TAVERNARAKIS N. Cellular stress response pathways and ageing: intricate molecular relationships[J]. The EMBO journal, 2011, 30(13): 2520-2531.

[4] BADYAEV A V. Stress-induced variation in evolution: from behavioural plasticity to genetic assimilation[J]. Proceedings of the royal society B: biological sciences, 2005, 272(1566): 877-886.

[5] LIU Y, CHANG A. Heat shock response relieves ER stress[J]. The EMBO journal, 2008, 27(7): 1049-1059.

[6] HAYNES C M, RON D. The mitochondrial UPR-protecting organelle protein homeostasis[J]. Journal of cell science, 2010, 123(22): 3849-3855.

[7] KIRSTEIN-MILES J, MORIMOTO R I. Caenorhabditis elegans as a model system to study intercompartmental proteostasis: interrelation of mitochondrial function, longevity, and neurodegenerative diseases[J]. Developmental dynamics, 2010, 239(5): 1529-1538.

[8] RON D, WALTER P. Signal integration in the endoplasmic reticulum unfolded protein response[J]. Nature reviews molecular cell biology, 2007, 8(7): 519-529.

[9] ÅKERFELT M, MORIMOTO R I, SISTONEN L. Heat shock factors: integrators of cell stress, development and lifespan[J]. Nature reviews molecular cell biology, 2010, 11(8): 545-555.

[10] SENFT D, RONAI Z A. UPR, autophagy, and mitochondria crosstalk un-

derlies the ER stress response[J]. Trends in biochemical sciences, 2015, 40(3):141-148.

[11] MOTTIS A, HERZIG S, AUWERX J. Mitocellular communication: shaping health and disease[J]. Science, 2019, 366(6467):827-832.

[12] HUGHES A L, GOTTSCHLING D E. An early age increase in vacuolar pH limits mitochondrial function and lifespan in yeast[J]. Nature, 2012, 492(7428):261-265.

[13] VEATCH J R, MCMURRAY M A, NELSON Z W, et al. Mitochondrial dysfunction leads to nuclear genome instability via an iron-sulfur cluster defect[J]. Cell, 2009, 137(7):1247-1258.

[14] HU F, LIU F. Mitochondrial stress: a bridge between mitochondrial dysfunction and metabolic diseases?[J]. Cellular signalling, 2011, 23(10):1528-1533.

[15] MOLENAARS M, JANSSENS G E, WILLIAMS E G, et al. A conserved mito-cytosolic translational balance links two longevity pathways[J]. Cell metabolism, 2020, 31(3):549-563.

[16] WANG X W, CHEN X J. A cytosolic network suppressing mitochondria-mediated proteostatic stress and cell death[J]. Nature, 2015, 524(7566):481-484.

[17] WROBEL L, TOPF U, BRAGOSZEWSKI P, et al. Mistargeted mitochondrial proteins activate a proteostatic response in the cytosol[J]. Nature, 2015, 524(7566):485-488.

[18] D'AMICO D, SORRENTINO V, AUWERX J. Cytosolic proteostasis networks of the mitochondrial stress response[J]. Trends in biochemical sciences, 2017, 42(9):712-725.

[19] FAKRUDDIN M, WEI F Y, SUZUKI T, et al. Defective mitochondrial tRNA taurine modification activates global proteostress and leads to mitochondrial disease[J]. Cell reports, 2018, 22(2):482-496.

[20] KIM H E, GRANT A R, SIMIC M S, et al. Lipid biosynthesis coordinates a mitochondrial-to-cytosolic stress response[J]. Cell, 2016, 166(6):1539-1552.

[21] HETZ C. The unfolded protein response: controlling cell fate decisions

under ER stress and beyond[J]. Nature reviews molecular cell biology, 2012,13(2):89-102.

[22] VANNUVEL K,RENARD P,RAES M,et al. Functional and morphological impact of ER stress on mitochondria[J]. Journal of cellular physiology, 2013,228(9):1802-1818.

[23] DEEGAN S,SAVELJEVA S,GORMAN A M,et al. Stress-induced self-cannibalism:on the regulation of autophagy by endoplasmic reticulum stress[J]. Cellular and molecular life sciences,2013,70(14):2425-2441.

[24] B'CHIR W,MAURIN A C,CARRARO V,et al. The eIF2α/ATF4 pathway is essential for stress-induced autophagy gene expression[J]. Nucleic acids research,2013,41(16):7683-7699.

[25] INDIVERI C,IACOBAZZI V,TONAZZI A,et al. The mitochondrial carnitine/acylcarnitine carrier:function,structure and physiopathology[J]. Molecular aspects of medicine,2011,32(4/5/6):223-233.

[26] JOVAISAITE V,MOUCHIROUD L,AUWERX J. The mitochondrial unfolded protein response,a conserved stress response pathway with implications in health and disease[J]. Journal of experimental biology,2014, 217(1):137-143.

[27] BRENNER S. The worm's turn[J]. Current biology,2009,95(4):6-7.

[28] BRENNER S. The genetics of caenorhabditis elegans[J]. Genetics,1974, 77(1):71-94

[29] FELIX M A. RNA interference in nematodes and the chance that favored Sydney Brenner[J]. Journal of biology,2008,7(9):34.

[30] FEINBERG E H,VANHOVEN M K,BENDESKY A,et al. GFP reconstitution across synaptic partners(GRASP) defines cell contacts and synapses in living nervous systems[J]. Neuron,2008,57(3):353-363.

[31] BOULIN T,ETCHBERGER J F,HOBERT O. Reporter gene fusions[J]. WormBook,2006,4(5):1-23.

[32] CHALFIE M,TU Y,EUSKIRCHEN G,et al. Green fluorescent protein as a marker for gene expression[J]. Science,1994,263(5148):802-805.

[33] KERR R A. Imaging the activity of neurons and muscles[J]. WormBook,2006,6(2):1-13.

[34] KIMBLE J, HIRSH D. The postembryonic cell lineages of the hermaphrodite and male gonads in Caenorhabditis elegans[J]. Developmantal biology,1979,70(2):396-417.

[35] SULSTON J E, HORVITZ H R. Post-embryonic cell lineages of the nematode, Caenorhabditis elegans[J]. Developmental biology,1977,56(1):110-156.

[36] SULSTON J E, SCHIERENBERG E, WHITE J G, et al. The embryonic cell lineage of the nematode Caenorhabditis elegans[J]. Developmental biology,1983,100(1):64-119.

[37] STIERNAGLE T. Maintenance of C. elegans[J]. WormBook,2006,11:1-11.

[38] RAIZEN D M, ZIMMERMAN J E, MAYCOCK M H, et al. Lethargus is a Caenorhabditis elegans sleep-like state[J]. Nature,2008,451(7178):569-572.

[39] ANKENY R A. The natural history of Caenorhabditis elegans research[J]. Nature reviews genetics,2001,2(6):474-479.

[40] KALETTA T, HENGARTNER M O. Finding function in novel targets: C. elegans as a model organism[J]. Nature reviews drug discovery,2006,5(5):387-398.

[41] CULETTO E, SATTELLE D B. A role for Caenorhabditis elegans in understanding the function and interactions of human disease genes[J]. Human molecular genetics,2000,9(6):869-877.

[42] SMITH M H, PLOEGH H L, WEISSMAN J S. Road to ruin: targeting proteins for degradation in the endoplasmic reticulum[J]. Science,2011,334(6059):1086-1090.

[43] FUJITA E, KOUROKU Y, ISOAI A, et al. Two endoplasmic reticulum-associated degradation (ERAD) systems for the novel variant of the mutant dysferlin: ubiquitin/proteasome ERAD(Ⅰ) and autophagy/lysosome ERAD(Ⅱ)[J]. Human molecular genetics,2007,16(6):618-629.

[44] BUKAU B, WEISSMAN J, HORWICH A. Molecular chaperones and protein quality control[J]. Cell,2006,125(3):443-451.

[45] KAUFMAN R J. Stress signaling from the lumen of the endoplasmic re-

ticulum: coordination of gene transcriptional and translational controls[J]. Genes & development,1999,13(10):1211-1233.

[46] MORI K. Tripartite management of unfolded proteins in the endoplasmic reticulum[J]. Cell,2000,101(5):451-454.

[47] URANO F,BERTOLOTTI A,RON D. IRE1 and efferent signaling from the endoplasmic reticulum[J]. Journal of cell science,2000,113(21):3697-3702.

[48] PATIL C,WALTER P. Intracellular signaling from the endoplasmic reticulum to the nucleus:the unfolded protein response in yeast and mammals[J]. Current opinion in cell biology,2001,13(3):349-355.

[49] SHIU R P,POUYSSEGUR J,PASTAN I. Glucose depletion accounts for the induction of two transformation-sensitive membrane proteins in Rous sarcoma virus-transformed chick embryo fibroblasts[J]. Proceedings of the national academy of sciences of the United States of America,1977,74(9):3840-3844.

[50] KOZUTSUMI Y,SEGAL M,NORMINGTON K,et al. The presence of malfolded proteins in the endoplasmic reticulum signals the induction of glucose-regulated proteins[J]. Nature,1988,332(6163):462-464.

[51] COX J S,SHAMU C E,WALTER P. Transcriptional induction of genes encoding endoplasmic reticulum resident proteins requires a transmembrane protein kinase[J]. Cell,1993,73(6):1197-1206.

[52] COX J S,WALTER P. A novel mechanism for regulating activity of a transcription factor that controls the unfolded protein response[J]. Cell,1996,87(3):391-404.

[53] MORL K,MA W,GETHING M-J,et al. A transmembrane protein with a cdc2+CDC28-related kinase activity is required for signaling from the ER to the nucleus[J]. Cell,1993,74(4):743-756.

[54] MORI K,KAWAHARA T,YOSHIDA H,et al. Signalling from endoplasmic reticulum to nucleus:transcription factor with a basic-leucine zipper motif is required for the unfolded protein-response pathway[J]. Genes to cells,1996,1(9):803-817.

[55] KAWAHARA T,YANAGI H,YURA T,et al. Endoplasmic reticulum

stress-induced mRNA splicing permits synthesis of transcription factor Hac1p/Ern4p that activates the unfolded protein response[J]. Molecular biology of the cell,1997,8(10):1845-1862.

[56] BERNALES S,PAPA F R,WALTER P. Intracellular signaling by the unfolded protein response[J]. Annual review of cell and developmental biology,2006,22(1):487-508.

[57] CALFON M,ZENG H,URANO F,et al. IRE1 couples endoplasmic reticulum load to secretory capacity by processing the XBP-1 mRNA[J]. Nature,2002,415(6867):92-96.

[58] SHEN X,ELLIS R E,LEE K,et al. Complementary signaling pathways regulate the unfolded protein response and are required for C. elegans development[J]. Cell,2001,107(7):893-903.

[59] YOSHIDA H,MATSUI T,YAMAMOTO A,et al. XBP1 mRNA is induced by ATF6 and spliced by IRE1 in response to ER stress to produce a highly active transcription factor[J]. Cell,2001,107(7):881-891.

[60] MORI K. Signalling pathways in the unfolded protein response:development from yeast to mammals[J]. Journal of biochemistry,2009,146(6):743-750.

[61] TIRASOPHON W,WELIHINDA A A,KAUFMAN R J. A stress response pathway from the endoplasmic reticulum to the nucleus requires a novel bifunctional protein kinase/endoribonuclease(Ire1p) in mammalian cells[J]. Genes & development,1998,12(12):1812-1824.

[62] BERTOLOTTI A,WANG X,NOVOA I,et al. Increased sensitivity to dextran sodium sulfate colitis in IRE1beta-deficient mice[J]. Journal of clinical in vestigation,2001,107(5):585-593.

[63] DARLING N J,COOK S J. The role of MAPK signalling pathways in the response to endoplasmic reticulum stress[J]. Biochimica et biophysica acta,2014,1843(10):2150-2163.

[64] SIDRAUSKI C,WALTER P. The transmembrane kinase Ire1p is a site-specific endonuclease that initiates mRNA splicing in the unfolded protein response[J]. Cell,1997,90(6):1031-1039.

[65] SHEN X,ELLIS R E,SAKAKI K,et al. Genetic interactions due to con-

stitutive and inducible gene regulation mediated by the unfolded protein response in C. elegans[J]. PLoS genetics,2005,1(3):e37.

[66] ACOSTA-ALVEAR D,ZHOU Y,BLAIS A,et al. XBP1 controls diverse cell type- and condition-specific transcriptional regulatory networks[J]. Molecular cell,2007,27(1):53-66.

[67] BEN-ZVI A,MILLER E A,MORIMOTO R I. Collapse of proteostasis represents an early molecular event in Caenorhabditis elegans aging[J]. Proceedings of the national academy of sciences of the United States of America,2009,106(35):14914-14919.

[68] HENIS-KORENBLIT S,ZHANG P,HANSEN M,et al. Insulin/IGF-1 signaling mutants reprogram ER stress response regulators to promote longevity[J]. Proceedings of the national academy of sciences of the United States of America,2010,107(21):9730-9735.

[69] TAYLOR R C,DILLIN A. XBP-1 is a cell-nonautonomous regulator of stress resistance and longevity[J]. Cell,2013,153(7):1435-1447.

[70] TAM A B,KOONG A C,NIWA M. Ire1 has distinct catalytic mechanisms for XBP1/HAC1 splicing and RIDD[J]. Cell reports,2014,9(3):850-858.

[71] GU F,NGUYEN D T,STUIBLE M,et al. Protein-tyrosine phosphatase 1B potentiates IRE1 signaling during endoplasmic reticulum stress[J]. Journal of biological chemistry,2004,279(48):49689-49693.

[72] GUPTA S,DEEPTI A,DEEGAN S,et al. HSP72 protects cells from ER stress-induced apoptosis via enhancement of IRE1alpha-XBP1 signaling through a physical interaction[J]. PLoS biology,2010,8(7):1-15.

[73] MARCU M G,DOYLE M,BERTOLOTTI A,et al. Heat shock protein 90 modulates the unfolded protein response by stabilizing IRE1alpha[J]. Molecular and cellular biology,2002,22(24):8506-8513.

[74] LEE A H,IWAKOSHI N N,GLIMCHER L H. XBP-1 regulates a subset of endoplasmic reticulum resident chaperone genes in the unfolded protein response[J]. Molecular and cellular biology,2003,23(21):7448-7459.

[75] SHAFFER A L,SHAPIRO-SHELEF M,IWAKOSHI N N,et al. XBP1,

downstream of Blimp-1, expands the secretory apparatus and other organelles, and increases protein synthesis in plasma cell differentiation[J]. Immunity, 2004, 21(1):81-93.

[76] SRIBURI R, BOMMIASAMY H, BULDAK G L, et al. Coordinate regulation of phospholipid biosynthesis and secretory pathway gene expression in XBP-1(S)-induced endoplasmic reticulum biogenesis[J]. Journal of biology chemistry, 2007, 282(10):7024-7034.

[77] RON D. Translational control in the endoplasmic reticulum stress response[J]. Journal of clinical investigation, 2002, 110(10):1383-1388.

[78] SHI Y, VATTEM K M, SOOD R, et al. Identification and characterization of pancreatic eukaryotic initiation factor 2 alpha-subunit kinase, PEK, involved in translational control[J]. Molecular and cellular biology, 1998, 18(12):7499-7509.

[79] HARDING H P, ZHANG Y, RON D. Protein translation and folding are coupled by an endoplasmic-reticulum-resident kinase[J]. Nature, 1999, 397(6716):271-274.

[80] BERTOLOTTI A, ZHANG Y, HENDERSHOT L M, et al. Dynamic interaction of BiP and ER stress transducers in the unfolded-protein response[J]. Nature cell biology, 2000, 2(6):326-332.

[81] MARCINIAK S J, GARCIA-BONILLA L, HU J, et al. Activation-dependent substrate recruitment by the eukaryotic translation initiation factor 2 kinase PERK[J]. Journal of cell biology, 2006, 172(2):201-209.

[82] HARDING H P, ZHANG Y, BERTOLOTTI A, et al. Perk is essential for translational regulation and cell survival during the unfolded protein response[J]. Molecular cell, 2000, 5(5):897-904.

[83] ITOH K, WAKABAYASHI N, KATOH Y, et al. Keap1 represses nuclear activation of antioxidant responsive elements by Nrf2 through binding to the amino-terminal Neh2 domain[J]. Genes & development, 1999, 13(1):76-86.

[84] CULLINAN S B, ZHANG D, HANNINK M, et al. Nrf2 is a direct PERK substrate and effector of PERK-dependent cell survival[J]. Molecular and cellular biology, 2003, 23(20):7198-7209.

[85] CULLINAN S B, DIEHL J A. PERK-dependent activation of Nrf2 contributes to redox homeostasis and cell survival following endoplasmic reticulum stress[J]. Journal of biological chemistry, 2004, 279(19): 20108-20117.

[86] SCHEUNER D, SONG B, MCEWEN E, et al. Translational control is required for the unfolded protein response and in vivo glucose homeostasis[J]. Molecular cell, 2001, 7(6): 1165-1176.

[87] HARDING H P, NOVOA I, ZHANG Y, et al. Regulated translation initiation controls stress-induced gene expression in mammalian cells[J]. Molecular cell, 2000, 6(5): 1099-1108.

[88] HARDING H P, ZHANG Y, ZENG H, et al. An integrated stress response regulates amino acid metabolism and resistance to oxidative stress[J]. Molecular cell, 2003, 11(3): 619-633.

[89] FAWCETT T W, MARTINDALE J L, GUYTON K Z, et al. Complexes containing activating transcription factor (ATF)/cAMP-responsive-element-binding protein (CREB) interact with the CCAAT/enhancer-binding protein (C/EBP)-ATF composite site to regulate Gadd153 expression during the stress response[J]. The biochemical journal, 1999, 339(1): 135-141.

[90] BRUHAT A, JOUSSE C, WANG X Z, et al. Amino acid limitation induces expression of CHOP, a CCAAT/enhancer binding protein-related gene, at both transcriptional and post-transcriptional levels[J]. Journal of biological chemistry, 1997, 272(28): 17588-17593.

[91] LUETHY J D, FARGNOLI J, PARK J S, et al. Isolation and characterization of the hamster gadd153 gene. Activation of promoter activity by agents that damage DNA[J]. Journal of biological chemistry, 1990, 265(27): 16521-16526.

[92] WANG X Z, LAWSON B, BREWER J W, et al. Signals from the stressed endoplasmic reticulum induce C/EBP-homologous protein (CHOP/GADD153)[J]. Molecular and cellular biology, 1996, 16(8): 4273-4280.

[93] RON D, HABENER J F. CHOP, a novel developmentally regulated nuclear protein that dimerizes with transcription factors C/EBP and LAP and functions as a dominant-negative inhibitor of gene transcription[J].

Genes & development, 1992, 6(3): 439-453.

[94] CHEN B P, WOLFGANG C D, HAI T. Analysis of ATF3, a transcription factor induced by physiological stresses and modulated by gadd153/Chop10[J]. Molecular and cellular biology, 1996, 16(3): 1157-1168.

[95] YAMAGUCHI H, WANG H G. CHOP is involved in endoplasmic reticulum stress-induced apoptosis by enhancing DR5 expression in human carcinoma cells[J]. Journal of biological chemistry, 2004, 279(44): 45495-45502.

[96] GHOSH A P, KLOCKE B J, BALLESTAS M E, et al. CHOP potentially co-operates with FOXO3a in neuronal cells to regulate PUMA and BIM expression in response to ER stress[J]. PLoS one, 2012, 7(6): e39586-1-e39586-11.

[97] PUTHALAKATH H, O'REILLY L A, GUNN P, et al. ER stress triggers apoptosis by activating BH3-only protein Bim[J]. Cell, 2007, 129(7): 1337-1349.

[98] MCCULLOUGH K D, MARTINDALE J L, KLOTZ L O, et al. Gadd153 sensitizes cells to endoplasmic reticulum stress by down-regulating Bcl2 and perturbing the cellular redox state[J]. Molecular and cellular biology, 2001, 21(4): 1249-1259.

[99] MARCINIAK S J, YUN C Y, OYADOMARI S, et al. CHOP induces death by promoting protein synthesis and oxidation in the stressed endoplasmic reticulum[J]. Genes & development, 2004, 18(24): 3066-3077.

[100] SONG B, SCHEUNER D, RON D, et al. Chop deletion reduces oxidative stress, improves beta cell function, and promotes cell survival in multiple mouse models of diabetes[J]. The journal of clinical investigation, 2008, 118(10): 3378-3389.

[101] LI G, MONGILLO M, CHIN K T, et al. Role of ERO1-alpha-mediated stimulation of inositol 1,4,5-triphosphate receptor activity in endoplasmic reticulum stress-induced apoptosis[J]. Journal of cell biology, 2009, 186(6): 783-792.

[102] HIGO T, HATTORI M, NAKAMURA T, et al. Subtype-specific and ER lumenal environment-dependent regulation of inositol 1,4,5-trisphos-

phate receptor type 1 by ERp44[J]. Cell,2005,120(1):85-98.

[103] KIVILUOTO S,VERVLIET T,IVANOVA H,et al. Regulation of inositol 1,4,5-trisphosphate receptors during endoplasmic reticulum stress [J]. Biochimica et biophysica acta,2013,1833(7):1612-1624.

[104] HAN J,BACK S H,HUR J,et al. ER-stress-induced transcriptional regulation increases protein synthesis leading to cell death[J]. Nature cell biology,2013,15(5):481-490.

[105] MA Y,HENDERSHOT L M. Delineation of a negative feedback regulatory loop that controls protein translation during endoplasmic reticulum stress[J]. Journal of biological chemistry, 2003, 278(37): 34864-34873.

[106] NOVOA I,ZENG H,HARDING H P,et al. Feedback inhibition of the unfolded protein response by GADD34-mediated dephosphorylation of eIF2alpha[J]. Journal of cell biology,2001,153(5):1011-1022.

[107] MORI K. Frame switch splicing and regulated intramembrane proteolysis:key words to understand the unfolded protein response[J]. Traffic, 2003,4(8):519-528.

[108] ADACHI Y,YAMAMOTO K,OKADA T,et al. ATF6 is a transcription factor specializing in the regulation of quality control proteins in the endoplasmic reticulum[J]. Cell structure and function,2008,33(1):75-89.

[109] YAMAMOTO K,SATO T,MATSUI T,et al. Transcriptional induction of mammalian ER quality control proteins is mediated by single or combined action of ATF6alpha and XBP1 [J]. Developmental cell, 2007,13(3):365-376.

[110] THUERAUF D J,MORRISON L,GLEMBOTSKI C C. Opposing roles for ATF6alpha and ATF6beta in endoplasmic reticulum stress response gene induction[J]. Journal of biological chemistry, 2004, 279 (20):21078-21084.

[111] HAZE K,OKADA T,YOSHIDA H,et al. Identification of the G13 (cAMP-response-element-binding protein-related protein) gene product related to activating transcription factor 6 as a transcriptional activator

of the mammalian unfolded protein response[J]. The biochemical journal,2001,355(1):19-28.

[112] HAZE K,YOSHIDA H,YANAGI H,et al. Mammalian transcription factor ATF6 is synthesized as a transmembrane protein and activated by proteolysis in response to endoplasmic reticulum stress[J]. Molecular biology of the cell,1999,10(11):3787-3799.

[113] NADANAKA S,YOSHIDA H,KANO F,et al. Activation of mammalian unfolded protein response is compatible with the quality control system operating in the endoplasmic reticulum[J]. Molecular biology of the cell,2004,15(6):2537-2548.

[114] SHEN J,CHEN X,HENDERSHOT L,et al. ER stress regulation of ATF6 localization by dissociation of BiP/GRP78 binding and unmasking of golgi localization signals[J]. Developmental cell,2002,3(1):99-111.

[115] YE J,RAWSON R B,KOMURO R,et al. ER stress induces cleavage of membrane-bound ATF6 by the same proteases that process SREBPs[J]. Molecular cell,2000,6(6):1355-1364.

[116] YOSHIDA H,OKADA T,HAZE K,et al. Endoplasmic reticulum stress-induced formation of transcription factor complex ERSF including NF-Y(CBF) and activating transcription factors 6alpha and 6beta that activates the mammalian unfolded protein response[J]. Molecular and cellular biology,2001,21(4):1239-1248.

[117] YOSHIDA H,OKADA T,HAZE K,et al. ATF6 activated by proteolysis binds in the presence of NF-Y(CBF) directly to the cis-acting element responsible for the mammalian unfolded protein response[J]. Molecular and cellular biology,2000,20(18):6755-6767.

[118] GETHING M J. Role and regulation of the ER chaperone BiP[J]. Seminars in cell & developmental biology,1999,10(5):465-472.

[119] ELLGAARD L,HELENIUS A. Quality control in the endoplasmic reticulum[J]. Nature reviews molecular cell biology,2003,4(3):181-191.

[120] SOMMER T,JAROSCH E. BiP binding keeps ATF6 at bay[J]. Developmental cell,2002,3(1):1-2.

[121] ROY B, LEE A S. The mammalian endoplasmic reticulum stress response element consists of an evolutionarily conserved tripartite structure and interacts with a novel stress-inducible complex[J]. Nucleic acids research,1999,27(6):1437-1443.

[122] YOSHIDA H, HAZE K, YANAGI H, et al. Identification of the cis-acting endoplasmic reticulum stress response element responsible for transcriptional induction of mammalian glucose-regulated proteins. Involvement of basic leucine zipper transcription factors[J]. Journal of biological chemistry,1998,273(50):33741-33749.

[123] LEE K, TIRASOPHON W, SHEN X, et al. IRE1-mediated unconventional mRNA splicing and S2P-mediated ATF6 cleavage merge to regulate XBP1 in signaling the unfolded protein response[J]. Genes & development,2002,16(4):452-466.

[124] LIN J H, WALTER P, YEN T S. Endoplasmic reticulum stress in disease pathogenesis[J]. Annual review of pathology,2008,3:399-425.

[125] LINDHOLM D, WOOTZ H, KORHONEN L. ER stress and neurodegenerative diseases[J]. Cell death and differentitation,2006,13(3):385-392.

[126] YOSHIDA H. ER stress and diseases[J]. FEBS journal,2007,274(3):630-658.

[127] FRANCISCO A B, SINGH R, SHA H, et al. Haploid insufficiency of suppressor enhancer Lin12 1-like(SEL1L) protein predisposes mice to high fat diet-induced hyperglycemia[J]. Journal of biological chemistry,2011,286(25):22275-22282.

[128] OZCAN U, CAO Q, YILMAZ E, et al. Endoplasmic reticulum stress links obesity, insulin action, and type 2 diabetes[J]. Science,2004,306(5695):457-461.

[129] KASER A, LEE A H, FRANKE A, et al. XBP1 links ER stress to intestinal inflammation and confers genetic risk for human inflammatory bowel disease[J]. Cell,2008,134(5):743-756.

[130] ZHANG K, WONG H N, SONG B, et al. The unfolded protein response sensor IRE1alpha is required at 2 distinct steps in B cell lym-

phopoiesis[J]. Journal of clinical investigation, 2005, 115(2):268-281.

[131] BISCHOF L J, KAO C Y, LOS F C, et al. Activation of the unfolded protein response is required for defenses against bacterial poreforming toxin in vivo[J]. PLoS pathogens, 2008, 4(10):e1000176-1-e1000176-11.

[132] SPRINGER W, HOPPE T, SCHMIDT E, et al. A Caenorhabditis elegans Parkin mutant with altered solubility couples alpha-synuclein aggregation to proteotoxic stress[J]. Human molecular genetics, 2005, 14(22):3407-3423.

[133] BURKEWITZ K, FENG G, DUTTA S, et al. Atf-6 regulates lifespan through ER-mitochondrial calcium homeostasis[J]. Cell reports, 2020, 32(10):108125.

[134] KENYON C J. The genetics of ageing[J]. Nature, 2010, 464(7288):504-512.

[135] LIN K, HSIN H, LIBINA N, et al. Regulation of the Caenorhabditis elegans longevity protein DAF-16 by insulin/IGF-1 and germline signaling[J]. Nature genetics, 2001, 28(2):139-145.

[136] JIA K, THOMAS C, AKBAR M, et al. Autophagy genes protect against Salmonella typhimurium infection and mediate insulin signaling-regulated pathogen resistance[J]. Proceedings of the national academy of sciences of the United States of America, 2009, 106(34):14564-14569.

[137] SAFRA M, FICKENTSCHER R, LEVI-FERBER M, et al. The FOXO transcription factor DAF-16 bypasses ire-1 requirement to promote endoplasmic reticulum homeostasis[J]. Cell metabolism, 2014, 20(5):870-881.

[138] ALBERT P S, BROWN S J, RIDDLE D L. Sensory control of dauer larva formation in Caenorhabditis elegans[J]. Journal of comparative neurology, 1981, 198(3):435-451.

[139] KIMURA K D, TISSENBAUM H A, LIU Y, et al. daf-2, an insulin receptor-like gene that regulates longevity and diapause in Caenorhabditis elegans[J]. Science, 1997, 277(5328):942-946.

[140] MORRIS J Z, TISSENBAUM H A, RUVKUN G. A phosphatidylinositol-3-OH kinase family member regulating longevity and diapause in Cae-

norhabditis elegans[J]. Nature,1996,382(6591):536-539.

[141] LIN K,DORMAN J B,RODAN A,et al. daf-16: An HNF-3/forkhead family member that can function to double the life-span of Caenorhabditis elegans[J]. Science,1997,278(5341):1319-1322.

[142] OGG S,PARADIS S,GOTTLIEB S,et al. The fork head transcription factor DAF-16 transduces insulin-like metabolic and longevity signals in C. elegans[J]. Nature,1997,389(6654):994-999.

[143] VOWELS J J,THOMAS J H. Genetic analysis of chemosensory control of dauer formation in Caenorhabditis elegans[J]. Genetics,1992,130(1):105-123.

[144] GOTTLIEB S,RUVKUN G. daf-2,daf-16 and daf-23:genetically interacting genes controlling Dauer formation in Caenorhabditis elegans[J]. Genetics,1994,137(1):107-120.

[145] OGG S,RUVKUN G. The C. elegans PTEN homolog,DAF-18,acts in the insulin receptor-like metabolic signaling pathway[J]. Molecular cell,1998,2(6):887-893.

[146] GIL E B,MALONE LINK E,LIU L X,et al. Regulation of the insulin-like developmental pathway of Caenorhabditis elegans by a homolog of the PTEN tumor suppressor gene[J]. Proceedings of the national academy of sciences of the United States of America,1999,96(6):2925-2930.

[147] ROUAULT J-P,KUWABARA P E,SINILNIKOVA O M,et al. Regulation of dauer larva development in Caenorhabditis elegans by daf-18, a homologue of the tumour suppressor PTEN[J]. Current biology,1999,9(6):329-334.

[148] MIHAYLOVA V T,BORLAND C Z,MANJARREZ L,et al. The PTEN tumor suppressor homolog in Caenorhabditis elegans regulates longevity and dauer formation in an insulin receptor-like signaling pathway[J]. Proceedings of the national academy of sciences of the United States of America,1999,96(13):7427-7432.

[149] KENYON C,CHANG J,GENSCH E,et al. A C. elegans mutant that lives twice as long as wild type[J]. Nature,1993,366(6454):461-464.

[150] KLASS M R. A method for the isolation of longevity mutants in the nematode Caenorhabditis elegans and initial results[J]. Mechanisms of ageing and development,1983,22(3/4):279-286.

[151] FRIEDMAN D B,JOHNSON T E. A mutation in the age-1 gene in Caenorhabditis elegans lengthens life and reduces hermaphrodite fertility[J]. Genetics,1988,118(1):75-86.

[152] FRIEDMAN D B, JOHNSON T E. Three mutants that extend both mean and maximum life span of the nematode,Caenorhabditis elegans, define the age-1 gene[J]. Journal of gerontology,1988,43(4):B102-B109.

[153] JOHNSON T E. Increased life-span of age-1 mutants in Caenorhabditis elegans and lower Gompertz rate of aging[J]. Science,1990,249(4971):908-912.

[154] WALKER G A, WHITE T M, MCCOLL G, et al. Heat shock protein accumulation is upregulated in a long-lived mutant of Caenorhabditis elegans[J]. The journals of gerontology series A-biological sciences and medical sciences,2001,56(7):B281-B287.

[155] LITHGOW G J, WHITE T M, MELOV S, et al. Thermotolerance and extended life-span conferred by single-gene mutations and induced by thermal stress[J]. Proceedings of the national academy of sciences of the United States of America,1995,92(16):7540-7544.

[156] HONDA Y,HONDA S. The daf-2 gene network for longevity regulates oxidative stress resistance and Mn-superoxide dismutase gene expression in Caenorhabditis elegans[J]. FASEB journal,1999,13(11):1385-1393.

[157] HONDA Y,HONDA S. Oxidative stress and life span determination in the nematode Caenorhabditis elegans[J]. Annals of the New York academy of sciences,2002,959:466-474.

[158] SCOTT B A,AVIDAN M S,CROWDER C M. Regulation of hypoxic death in C. elegans by the insulin/IGF receptor homolog DAF-2[J]. Science,2002,296(5577):2388-2391.

[159] LAMITINA S T,STRANGE K. Transcriptional targets of DAF-16 insu-

lin signaling pathway protect C. elegans from extreme hypertonic stress[J]. American journal of physiology-cell physiology, 2005, 288(2): C467-C474.

[160] BARSYTE D, LOVEJOY D A, LITHGOW G J. Longevity and heavy metal resistance in daf-2 and age-1 long-lived mutants of Caenorhabditis elegans[J]. FASEB journal, 2001, 15(3): 627-634.

[161] MURAKAMI S, JOHNSON T E. A genetic pathway conferring life extension and resistance to UV stress in Caenorhabditis elegans[J]. Genetics, 1996, 143(3): 1207-1218.

[162] TEIXEIRA-CASTRO A, AILION M, JALLES A, et al. Neuron-specific proteotoxicity of mutant ataxin-3 in C. elegans: rescue by the DAF-16 and HSF-1 pathways[J]. Human molecular genetics, 2011, 20(15): 2996-3009.

[163] HSU A L, MURPHY C T, KENYON C. Regulation of aging and age-related disease by DAF-16 and heat-shock factor[J]. Science, 2003, 300(5622): 1142-1145.

[164] COHEN E, BIESCHKE J, PERCIAVALLE R M, et al. Opposing activities protect against age-onset proteotoxicity[J]. Science, 2006, 313(5793): 1604-1610.

[165] ASHRAFI K, CHANG F Y, WATTS J L, et al. Genome-wide RNAi analysis of Caenorhabditis elegans fat regulatory genes[J]. Nature, 2003, 421(6920): 268-272.

[166] PEREZ C L, VAN GILST M R. A 13C isotope labeling strategy reveals the influence of insulin signaling on lipogenesis in C. elegans[J]. Cell metabolism, 2008, 8(3): 266-274.

[167] O'ROURKE E J, SOUKAS A A, CARR C E, et al. C. elegans major fats are stored in vesicles distinct from lysosome-related organelles[J]. Cell metabolism, 2009, 10(5): 430-435.

[168] GARSIN D A, VILLANUEVA J M, BEGUN J, et al. Long-lived C. elegans daf-2 mutants are resistant to bacterial pathogens[J]. Science, 2003, 300(5627): 1921.

[169] KERRY S, TEKIPPE M, GADDIS N C, et al. GATA transcription fac-

tor required for immunity to bacterial and fungal pathogens[J]. PLoS one,2006,1:e77.

[170] HUGHES S E,EVASON K,XIONG C,et al. Genetic and pharmacological factors that influence reproductive aging in nematodes[J]. PLoS genetics,2007,3(2):254-265.

[171] LUO S,KLEEMANN G A,ASHRAF J M,et al. TGF-beta and insulin signaling regulate reproductive aging via oocyte and germline quality maintenance[J]. Cell,2010,143(2):299-312.

[172] LUO S,SHAW W M,ASHRAF J,et al. TGF-beta Sma/Mab signaling mutations uncouple reproductive aging from somatic aging[J]. PLoS genetics,2009,5(12):e1000789-1-e1000789-15.

[173] ULLRICH A,GRAY A,TAM A W,et al. Insulin-like growth factor I receptor primary structure: comparison with insulin receptor suggests structural determinants that define functional specificity[J]. EMBO journal,1986,5(10):2503-2512.

[174] ULLRICH A,BELL J R,CHEN E Y,et al. Human insulin receptor and its relationship to the tyrosine kinase family of oncogenes[J]. Nature,1985,313(6005):756-761.

[175] WHITE M F. The IRS-signalling system: a network of docking proteins that mediate insulin action[J]. Molecular and cellular biochemistry,1998,182:3-11.

[176] TANIGUCHI C M,EMANUELLI B,KAHN C R. Critical nodes in signalling pathways:insights into insulin action[J]. Nature reviews molecular cell biology,2006,7(2):85-96.

[177] PINERO GONZALEZ J,CARRILLO FARNES O,VASCONCELOS A T, et al. Conservation of key members in the course of the evolution of the insulin signaling pathway[J]. Biosystems,2009,95(1):7-16.

[178] HUA Q X,NAKAGAWA S H,WILKEN J,et al. A divergent INS protein in Caenorhabditis elegans structurally resembles human insulin and activates the human insulin receptor[J]. Genes & development, 2003,17(7):826-831.

[179] MURPHY C T,HU P J. Insulin/insulin-like growth factor signaling in

C. elegans[J]. WormBook 2009,12:1-43.

[180] TULLET J M, HERTWECK M, AN J H, et al. Direct inhibition of the longevity-promoting factor SKN-1 by insulin-like signaling in C. elegans [J]. Cell,2008,132(6):1025-1038.

[181] LEE R Y N, HENCH J, RUVKUN G. Regulation of C. elegans DAF-16 and its human ortholog FKHRL1 by the daf-2 insulin-like signaling pathway[J]. Current biology,2001,11(24):1950-1957.

[182] TRAN H, BRUNET A, GRIFFITH E C, et al. The many forks in FOXO's road[J]. Signal transduction knowledge environment,2003,2003(172):RE5.

[183] BRUNET A, PARK J, TRAN H, et al. Protein kinase SGK mediates survival signals by phosphorylating the forkhead transcription factor FKHRL1(FOXO3a)[J]. Molecular and cellular biology,2001,21(3):952-965.

[184] HERTWECK M, GÖBEL C, BAUMEISTER R. C. elegans SGK-1 is the critical component in the Akt/PKB kinase complex to control stress response and life span[J]. Developmental cell,2004,6(4):577-588.

[185] XIAO R, ZHANG B, DONG Y, et al. A genetic program promotes C. elegans longevity at cold temperatures via a thermosensitive TRP channel[J]. Cell,2013,152(4):806-817.

[186] DAVIS R J. Signal Transduction by the JNK group of MAP kinases [J]. Cell,2000,103(2):239-252.

[187] OH S W, MUKHOPADHYAY A, SVRZIKAPA N, et al. JNK regulates lifespan in Caenorhabditis elegans by modulating nuclear translocation of forkhead transcription factor/DAF-16[J]. Proceedings of the national academy of sciences of the United Stated of America,2005,102(12):4494-4499.

[188] WANG M C, BOHMANN D, JASPER H. JNK extends life span and limits growth by antagonizing cellular and organism-wide responses to insulin signaling[J]. Cell,2005,121(1):115-125.

[189] ESSERS M A, WEIJZEN S, DE VRIES-SMITS A M, et al. FOXO tran-

scription factor activation by oxidative stress mediated by the small GTPase Ral and JNK[J]. The EMBO journal,2004,23(24):4802-4812.

[190] GREER E L,DOWLATSHAHI D,BANKO M R,et al. An AMPK-FOXO pathway mediates longevity induced by a novel method of dietary restriction in C. elegans[J]. Current biology,2007,17(19):1646-1656.

[191] NARASIMHAN S D,YEN K,TISSENBAUM H A. Converging pathways in lifespan regulation[J]. Current biology, 2009, 19(15): R657-R666.

[192] JONES K T,GREER E R,PEARCE D,et al. Rictor/TORC2 regulates Caenorhabditis elegans fat storage, body size, and development through sgk-1[J]. PLoS biology,2009,7(3):e60.

[193] SOUKAS A A,KANE E A,CARR C E,et al. Rictor/TORC2 regulates fat metabolism, feeding, growth, and life span in Caenorhabditis elegans [J]. Genes & development,2009,23(4):496-511.

[194] ROBIDA-STUBBS S,GLOVER-CUTTER K,LAMMING D W,et al. TOR signaling and rapamycin influence longevity by regulating SKN-1/Nrf and DAF-16/FoxO[J]. Cell metabolism,2012,15(5):713-724.

[195] VOLOVIK Y,MOLL L,MARQUES F C,et al. Differential regulation of the heat shock factor 1 and DAF-16 by neuronal nhl-1 in the nematode C. elegans[J]. Cell reports,2014,9(6):2192-2205.

[196] RIEDEL C G,DOWEN R H,LOURENCO G F,et al. DAF-16/FOXO employs the chromatin remodeller SWI/SNF to promote stress resistance and longevity[J]. Nature cell biology,2013,15(5):491-501.

[197] BANSAL A,KWON E S,CONTE D,et al. Transcriptional regulation of Caenorhabditis elegans FOXO/DAF-16 modulates lifespan[J]. Longevity & healthspan,2014,3:5.

[198] CAHILL C M,TZIVION G,NASRIN N,et al. Phosphatidylinositol 3-kinase signaling inhibits DAF-16 DNA binding and function via 14-3-3-dependent and 14-3-3-independent pathways[J]. Journal of biology chemistry,2001,276(16):13402-13410.

[199] BURGERING B M T,KOPS G J P L. Cell cycle and death control:

long live forkheads[J]. Trends in biochemical sciences,2002,27(7): 352-360.

[200] BURGERING B M T. A brief introduction to FOXOlogy[J]. Oncogene, 2008,27(16):2258-2262.

[201] BERDICHEVSKY A,VISWANATHAN M,HORVITZ H R,et al. C. elegans SIR-2.1 interacts with 14-3-3 proteins to activate DAF-16 and extend life span[J]. Cell,2006,125(6):1165-1177.

[202] LI J,TEWARI M,VIDAL M,et al. The 14-3-3 protein FTT-2 regulates DAF-16 in Caenorhabditis elegans[J]. Development biology,2007, 301(1):82-91.

[203] WANG Y,OH S W,DEPLANCKE B,et al. C. elegans 14-3-3 proteins regulate life span and interact with SIR-2.1 and DAF-16/FOXO[J]. Mechanisms of ageing & developrnent,2006,127(9):741-747.

[204] RINE J,HERSKOWITZ I. Four genes responsible for a position effect on expression from HML and HMR in saccharomyces cerevisiae[J]. Genetics,1987,116(1):9-22.

[205] KAEBERLEIN M,MCVEY M,GUARENTE L. The SIR2/3/4 complex and SIR2 alone promote longevity in Saccharomyces cerevisiae by two different mechanisms[J]. Genes & development, 1999, 13(19): 2570-2580.

[206] ROGINA B,HELFAND S L. Sir2 mediates longevity in the fly through a pathway related to calorie restriction[J]. Proceedings of the national academy of sciences of the United Stated of America,2004,101(45): 15998-16003.

[207] TISSENBAUM H A,GUARENTE L. Increased dosage of a sir-2 gene extends lifespan in Caenorhabditis elegans [J]. Nature, 2001, 410 (6825):227-230.

[208] DIEDERICH R J,MERRILL V K,PULTZ M A,et al. Isolation,structure, and expression of labial, a homeotic gene of the antennapedia complex involved in drosophila head development[J]. Genes & development,1989,3(3):399-414.

[209] WEIGEL D,JÜRGENS G,KÜTTNER F,et al. The homeotic gene fork

head encodes a nuclear protein and is expressed in the terminal regions of the drosophila embryo[J]. Cell,1989,57(4):645-658.

[210] REIS L M,TYLER R C,SCHNEIDER A,et al. FOXE3 plays a significant role in autosomal recessive microphthalmia[J]. American journal of medical genetics part A,2010,152A(3):582-590.

[211] DOUCETTE L,GREEN J,FERNANDEZ B,et al. A novel,non-stop mutation in FOXE3 causes an autosomal dominant form of variable anterior segment dysgenesis including Peters anomaly[J]. European journal of human genetics,2011,19(3):293-299.

[212] KRALL M,HTUN S,ANAND D,et al. A zebrafish model of foxe3 deficiency demonstrates lens and eye defects with dysregulation of key genes involved in cataract formation in humans[J]. Human genetics,2018,137(4):315-328.

[213] ENARD W,PRZEWORSKI M,FISHER S E,et al. Molecular evolution of FOXP2,a gene involved in speech and language[J]. Nature,2002,418(6900):869-872.

[214] MARCUS G F,FISHER S E. FOXP2 in focus:what can genes tell us about speech and language?[J]. Trends in cognitive sciences,2003,7(6):257-262.

[215] VARGHA-KHADEM F,GADIAN D G,COPP A,et al. FOXP2 and the neuroanatomy of speech and language[J]. Nature reviews neuroscience,2005,6(2):131-138.

[216] CARTER M E,BRUNET A. FOXO transcription factors[J]. Current biology,2007,17(4):R113-R114.

[217] GILLEY J,COFFER P J,HAM J. FOXO transcription factors directly activate bim gene expression and promote apoptosis in sympathetic neurons[J]. Journal of cell biology,2003,162(4):613-622.

[218] YANG J-Y,XIA W,HU M. Ionizing radiation activates expression of FOXO3a, Fas ligand, and Bim, and induces cell apoptosis[J]. International journal of oncology,2006,29(3):643-648.

[219] FARHAN M,WANG H,GAUR U,et al. FOXO signaling pathways as therapeutic targets in cancer[J]. Internationa journal of biological sci-

ences,2017,13(7):815-827.

[220] TRAN H,BRUNET A,GRENIER J M,et al. DNA repair pathway stimulated by the forkhead transcription factor FOXO3a through the Gadd45 protein[J]. Science,2002,296(5567):530-534.

[221] LI M,CHIU J F,MOSSMAN B T,et al. Down-regulation of manganese-superoxide dismutase through phosphorylation of FOXO3a by Akt in explanted vascular smooth muscle cells from old rats[J]. Journal of biological chemistry,2006,281(52):40429-40439.

[222] LIM S W,JIN L,LUO K,et al. Klotho enhances FoxO3-mediated manganese superoxide dismutase expression by negatively regulating PI3K/AKT pathway during tacrolimus-induced oxidative stress[J]. Cell death & disease,2017,8(8):e2972.

[223] TONG X,ZHANG D,CHARNEY N,et al. DDB1-mediated CRY1 degradation promotes FOXO1-driven gluconeogenesis in liver[J]. Diabetes,2017,66(10):2571-2582.

[224] CHEN Y R,LIU M T,CHANG Y T,et al. Epstein-Barr virus latent membrane protein 1 represses DNA repair through the PI3K/Akt/FOXO3a pathway in human epithelial cells[J]. Journal of virology,2008,82(16):8124-8137.

[225] HORNSVELD M,DANSEN T B,DERKSEN P W,et al. Re-evaluating the role of FOXOs in cancer[J]. Seminars in cancer biology,2018,50:90-100.

[226] HUANG H,TINDALL D J. Dynamic FoxO transcription factors[J]. Journal of cell science,2007,120(15):2479-2487.

[227] GIANNAKOU M E,PARTRIDGE L. The interaction between FOXO and SIRT1:tipping the balance towards survival[J]. Trends in cell biology,2004,14(8):408-412.

[228] HARIHARAN N,MAEJIMA Y,NAKAE J,et al. Deacetylation of FoxO by sirt1 plays an essential role in mediating starvation-induced autophagy in cardiac myocytes[J]. Circulation research,2010,107(12):1470-1482.

[229] WANG F,CHAN C H,CHEN K,et al. Deacetylation of FOXO3 by

SIRT1 or SIRT2 leads to Skp2-mediated FOXO3 ubiquitination and degradation[J]. Oncogene,2012,31(12):1546-1557.

[230] HUANG H,TINDALL D J. Regulation of FOXO protein stability via ubiquitination and proteasome degradation[J]. Biochimica et biophysica acta,2011,1813(11):1961-1964.

[231] SLACK C,GIANNAKOU M E,FOLEY A,et al. dFOXO-independent effects of reduced insulin-like signaling in drosophila[J]. Aging cell,2011,10(5):735-748.

[232] SHENG M,GREENBERG M E. The regulation and function of c-fos and other immediate early genes in the nervous system[J]. Neuron,1990,4(4):477-485.

[233] MEYER T E,HABENER J F. Cyclic adenosine 3′,5′-monophosphate response element binding protein(CREB) and related transcription-activating deoxyribonucleic acid-binding proteins[J]. Endocrine reviews,1993,14(3):269-290.

[234] LEE K A W,MASSON N. Transcriptional regulation by CREB and its relatives[J]. Biochimica et biophysica acta-gene structure and expression,1993,1174(3):221-233.

[235] OMORI Y,IMAI J,WATANABE M,et al. CREB-H:a novel mammalian transcription factor belonging to the CREB/ATF family and functioning via the box-B element with a liver-specific expression[J]. Nucleic acids research,2001,29(10):2154-2162.

[236] NAKAGAWA Y,SHIMANO H. CREBH regulates systemic glucose and lipid metabolism[J]. International journal of molecular sciences,2018,19(5):1396.

[237] CHIN K T,ZHOU H J,WONG C M,et al. The liver-enriched transcription factor CREB-H is a growth suppressor protein underexpressed in hepatocellular carcinoma[J]. Nucleic acids research,2005,33(6):1859-1873.

[238] CUI A,DING D,LI Y. Regulation of hepatic metabolism and cell growth by the ATF/CREB family of transcription factors[J]. Diabetes,2021,70(3):653-664.

[239] NAKAGAWA Y, SATOH A, YABE S, et al. Hepatic CREB3L3 controls whole-body energy homeostasis and improves obesity and diabetes [J]. Endocrinology, 2014, 155(12):4706-4719.

[240] NAKAGAWA Y, SATOH A, TEZUKA H, et al. CREB3L3 controls fatty acid oxidation and ketogenesis in synergy with PPAR α[J]. Scientific reports, 2016, 6:39182.

[241] KIKUCHI T, ORIHARA K, OIKAWA F, et al. Intestinal CREBH overexpression prevents high-cholesterol diet-induced hypercholesterolemia by reducing Npc1l1 expression[J]. Molecular metabolism, 2016, 5(11): 1092-1102.

[242] SILVA M C, FOX S, BEAM M, et al. A genetic screening strategy identifies novel regulators of the proteostasis network[J]. PLoS genetics, 2011, 7(12):e1002438-1-e1002438-15.

[243] GUISBERT E, CZYZ D M, RICHTER K, et al. Identification of a tissue-selective heat shock response regulatory network[J]. PLoS genetics, 2013, 9(4):e1003466.

[244] WEICKSEL S E, MAHADAV A, MOYLE M, et al. A novel small molecule that disrupts a key event during the oocyte-to-embryo transition in C. elegans[J]. Development, 2016, 143(19):3540-3548.

[245] HOUTKOOPER R H, ARGMANN C, HOUTEN S M, et al. The metabolic footprint of aging in mice[J]. Scientific reports, 2011, 1:134.

[246] LAWTON K A, BERGER A, MITCHELL M, et al. Analysis of the adult human plasma metabolome[J]. Pharmacogenomics, 2008, 9(4): 383-397.

[247] GAO A W, CHATZISPYROU I A, KAMBLE R, et al. A sensitive mass spectrometry platform identifies metabolic changes of life history traits in C. elegans[J]. Scientific reports, 2017, 7(1):2408.

[248] MOGHADAM N N, HOLMSTRUP M, PERTOLDI C, et al. Age-induced perturbation in cell membrane phospholipid fatty acid profile of longevity-selected Drosophila melanogaster and corresponding control lines [J]. Experimental gerontology, 2013, 48(11):1362-1368.

[249] HAHN O, GRONKE S, STUBBS T M, et al. Dietary restriction protects

from age-associated DNA methylation and induces epigenetic reprogramming of lipid metabolism[J]. Genome biology,2017,18(1):56.

[250] HANSEN M,FLATT T,AGUILANIU H. Reproduction,fat metabolism, and life span:what is the connection?[J]. Cell metabolism,2013,17(1):10-19.

[251] CLELLAND E,PENG C. Endocrine/paracrine control of zebrafish ovarian development[J]. Molecular and cellular endocrinology,2009,312(1/2):42-52.

[252] YAN Y-L,POSTLETHWAIT J H. Vitellogenesis in drosophila:sequestration of a yolk polypeptide/invertase fusion protein into developing oocytes[J]. Developmental biology,1990,140(2):281-290.

[253] KIMBLE J,SHARROCK W J. Tissue-specific synthesis of yolk proteins in Caenorhabditis elegans[J]. Developmental biology,1983,96(1):189-196.

[254] GONZALEZ-COVARRUBIAS V. Lipidomics in longevity and healthy aging[J]. Biogerontology,2013,14(6):663-672.

[255] HULBERT A J,PAMPLONA R,BUFFENSTEIN R,et al. Life and death:metabolic rate,membrane composition,and life span of animals[J]. Physiological reviews,2007,87(4):1175-1213.

[256] LIN L,CAO B,XU Z,et al. In vivo HMRS and lipidomic profiling reveals comprehensive changes of hippocampal metabolism during aging in mice[J]. Biochemical and biophysical research communications,2016,470(1):9-14.

[257] BRAUN F,RINSCHEN M M,BARTELS V,et al. Altered lipid metabolism in the aging kidney identified by three layered omic analysis[J]. Aging,2016,8(3):441-454.

[258] MITCHELL T W,BUFFENSTEIN R,HULBERT A J. Membrane phospholipid composition may contribute to exceptional longevity of the naked mole-rat(heterocephalus glaber):a comparative study using shotgun lipidomics[J]. Experimental gerontology,2007,42(11):1053-1062.

[259] YU B P,SUESCUN E A,YANG S Y. Effect of age-related lipid peroxidation on membrane fluidity and phospholipase A2:modulation by di-

etary restriction[J]. Mechanisms of ageing and development,1992,65(1):17-33.

[260] MONTOLIU I,SCHERER M,BEGUELIN F,et al. Serum profiling of healthy aging identifies phospho- and sphingolipid species as markers of human longevity[J]. Aging,2015,7(11):9-25.

[261] SUNSHINE H,IRUELA-ARISPE M L. Membrane lipids and cell signaling[J]. Current opinion in lipidology,2017,28(5):408-413.

[262] SEZGIN E,LEVENTAL I,MAYOR S,et al. The mystery of membrane organization:composition,regulation and roles of lipid rafts[J]. Nature reviews molecular cell biology,2017,18(6):361-374.

[263] PAPSDORF K,BRUNET A. Linking lipid metabolism to chromatin regulation in aging[J]. Trends in cell biology,2019,29(2):97-116.

[264] BROWN M S,GOLDSTEIN J L. The SREBP pathway:regulation of cholesterol metabolism by proteolysis of a membrane-bound transcription factor[J]. Cell,1997,89(3):331-340.

[265] BROWN M S,YE J,RAWSON R B,et al. Regulated intramembrane proteolysiss:a control mechanism conserved from bacteria to humans[J]. Cell,2000,100(4):391-398.

[266] SAMPIERI L,DI GIUSTO P,ALVAREZ C. CREB3 transcription factors:ER-Golgi stress transducers as hubs for cellular homeostasis[J]. Frontiers in cell and developmental biology,2019,7:123.

[267] ZHANG K,SHEN X,WU J,et al. Endoplasmic reticulum stress activates cleavage of CREBH to induce a systemic inflammatory response[J]. Cell,2006,124(3):587-599.

[268] ZHOU L,HE B,DENG J,et al. Histone acetylation promotes long-lasting defense responses and longevity following early life heat stress[J]. PLoS genetics,2019,15(4):e1008122.

[269] TROEMEL E R,CHU S W,REINKE V,et al. p38 MAPK regulates expression of immune response genes and contributes to longevity in C. elegans[J]. PLoS genetics,2006,2(11):e183-1-e183-15.

[270] YUNGER E,SAFRA M,LEVI-FERBER M,et al. Innate immunity mediated longevity and longevity induced by germ cell removal converge

on the C-type lectin domain protein IRG-7[J]. PLoS genetics, 2017, 13(2):e1006577.

[271] KUMAR S, EGAN B M, KOCSISOVA Z, et al. Lifespan extension in C. elegans caused by bacterial colonization of the intestine and subsequent activation of an innate immune response [J]. Developmental cell, 2019, 49(1):100-117.

[272] MARTINS R, LITHGOW G J, LINK W. Long live FOXO: unraveling the role of FOXO proteins in aging and longevity[J]. Aging cell, 2016, 15(2):196-207.

[273] ZOU L, WU D, ZANG X, et al. Construction of a germline-specific RNAi tool in C. elegans[J]. Scientific reports, 2019, 9(1):2354.

[274] KAMATH R S, FRASER A G, DONG Y, et al. Systematic functional analysis of the Caenorhabditis elegans genome using RNAi[J]. Nature, 2003, 421(6920):231-237.

[275] KIM D, PERTEA G, TRAPNELL C, et al. TopHat2: accurate alignment of transcriptomes in the presence of insertions, deletions and gene fusions[J]. Genome biology, 2013, 14(4):R36.

[276] TRAPNELL C, WILLIAMS B A, PERTEA G, et al. Transcript assembly and quantification by RNA-Seq reveals unannotated transcripts and isoform switching during cell differentiation[J]. Nature biotechnology, 2010, 28(5):511-515.

[277] TRAPNELL C, ROBERTS A, GOFF L, et al. Differential gene and transcript expression analysis of RNA-seq experiments with TopHat and Cufflinks[J]. Nature protocols, 2012, 7(3):562-578.

[278] TRAPNELL C, HENDRICKSON D G, SAUVAGEAU M, et al. Differential analysis of gene regulation at transcript resolution with RNA-seq[J]. Nature biotechnology, 2013, 31(1):46-53.

[279] DENNIS G, SHERMAN B T, HOSACK D A, et al. DAVID: database for annotation, visualization, and integrated discovery[J]. Genome biology, 2003, 4(5):P3.

[280] BOLGER A M, LOHSE M, USADEL B. Trimmomatic: a flexible trimmer for illumina sequence data[J]. Bioinformatics, 2014, 30(15):2114-

2120.

[281] LI H, DURBIN R. Fast and accurate short read alignment with Burrows-Wheeler transform[J]. Bioinformatics, 2010, 25(14): 1754-1760.

[282] LI H, HANDSAKER B, WYSOKER A, et al. The sequence alignment/map format and SAMtools[J]. Bioinformatics, 2009, 25(16): 2078-2079.

[283] SALMON-DIVON M, DVINGE H, TAMMOJA K, et al. PeakAnalyzer: genome-wide annotation of chromatin binding and modification loci[J]. BMC bioinformatics, 2010, 11(1): 1.

[284] WATTS J L, BROWSE J. Genetic dissection of polyunsaturated fatty acid synthesis in Caenorhabditis elegans[J]. Proceedings of the national academy of sciences of the United States of America, 2002, 99(9): 5854-5859.

[285] SHIVERS R P, KOOISTRA T, CHU S W, et al. Tissue-specific activities of an immune signaling module regulate physiological responses to pathogenic and nutritional bacteria in C. elegans[J]. Cell host microbe, 2009, 6(4): 321-330.

[286] KIM D H, FEINBAUM R, ALLOING G, et al. A conserved p38 MAP kinase pathway in Caenorhabditis elegans innate immunity[J]. Science, 2002, 297(5581): 623-626.

[287] SINGH V, ABALLAY A. Heat-shock transcription factor (HSF)-1 pathway required for Caenorhabditis elegans immunity[J]. Proceedings of the national academy of sciences of the United States of America, 2006, 103(35): 13092-13097.

[288] HOEVEN R, MCCALLUM K C, CRUZ M R, et al. Ce-Duox1/BLI-3 generated reactive oxygen species trigger protective SKN-1 activity via p38 MAPK signaling during infection in C. elegans[J]. PLoS pathogens, 2011, 7(12): e1002453.

[289] PAPP D, CSERMELY P, SOTI C. A role for SKN-1/Nrf in pathogen resistance and immunosenescence in Caenorhabditis elegans[J]. PLoS pathogens, 2012, 8(4): e1002673.

[290] REA S L, WU D, CYPSER J R, et al. A stress-sensitive reporter predicts longevity in isogenic populations of Caenorhabditis elegans[J].

Nature genetics, 2005, 37(8): 894-898.

[291] LEIERS B, KAMPKÖTTER A, GREVELDING C G, et al. A stress-responsive glutathione S-transferase confers resistance to oxidative stress in Caenorhabditis elegans[J]. Free radical biology and medicine, 2003, 34(11): 1405-1415.

[292] OGAMI K, CHEN Y, MANLEY J L. RNA surveillance by the nuclear RNA exosome: mechanisms and significance[J]. Noncoding RNA, 2018, 4(1): 8.

[293] BARBOSA S, FASANELLA G, CARREIRA S, et al. An orchestrated program regulating secretory pathway genes and cargos by the transmembrane transcription factor CREB-H[J]. Traffic, 2013, 14(4): 382-398.

[294] TEPPER R G, ASHRAF J, KALETSKY R, et al. PQM-1 complements DAF-16 as a key transcriptional regulator of DAF-2-mediated development and longevity[J]. Cell, 2013, 154(3): 676-690.

[295] BERMAN J R, KENYON C. Germ-cell loss extends C. elegans life span through regulation of DAF-16 by kri-1 and lipophilic-hormone signaling[J]. Cell, 2006, 124(5): 1055-1068.

[296] GERISCH B, WEITZEL C, KOBER-EISERMANN C, et al. A hormonal signaling pathway influencing C. elegans metabolism, reproductive development, and life span[J]. Developmental cell, 2001, 1(6): 841-851.

[297] GERISCH B, ROTTIERS V, LI D, et al. A bile acid-like steroid modulates Caenorhabditis elegans lifespan through nuclear receptor signaling[J]. Proceedings of the national academy of sciences of the United States of America, 2007, 104(12): 5014-5019.

[298] AMRIT F R, STEENKISTE E M, RATNAPPAN R, et al. DAF-16 and TCER-1 facilitate adaptation to germline loss by restoring lipid homeostasis and repressing reproductive physiology in C. elegans[J]. PLoS genetics, 2016, 12(10): e1006381.

[299] AMRIT F R G, NAIM N, RATNAPPAN R, et al. The longevity-promoting factor, TCER-1, widely represses stress resistance and innate immunity[J]. Nature communications, 2019, 10(11): 1-16.

[300] MILLER H, FLETCHER M, PRIMITIVO M, et al. Genetic interaction with temperature is an important determinant of nematode longevity [J]. Aging cell, 2017, 16(6): 1425-1429.

[301] KIM H, MENDEZ R, ZHENG Z, et al. Liver-enriched transcription factor CREBH interacts with peroxisome proliferator-activated receptor alpha to regulate metabolic hormone FGF21[J]. Endocrinology, 2014, 155 (3): 769-782.

[302] LEWIS K N, RUBINSTEIN N D, BUFFENSTEIN R. A window into extreme longevity: the circulating metabolomic signature of the naked mole-rat, a mammal that shows negligible senescence[J]. Geroscience, 2018, 40(2): 105-121.

[303] CHANG C L, HO M C, LEE P H, et al. S1P(5) is required for sphingosine 1-phosphate-induced autophagy in human prostate cancer PC-3 cells[J]. American journal of physiology, 2009, 297(2): C451-C458.

[304] HUANG Y-L, CHANG C-L, TANG C-H, et al. Extrinsic sphingosine 1-phosphate activates S1P5 and induces autophagy through generating endoplasmic reticulum stress in human prostate cancer PC-3 cells[J]. Cellular signalling, 2014, 26(3): 611-618.

[305] GREEN C, MITCHELL S, SPEAKMAN J. Energy balance and the sphingosine-1-phosphate/ceramide axis[J]. Aging, 2017, 9(12): 2463-2464.

[306] HUANG X, WITHERS B R, DICKSON R C. Sphingolipids and lifespan regulation[J]. Biochimica et biophysica acta, 2014, 1841(5): 657-664.

[307] CUTLER R G, THOMPSON K W, CAMANDOLA S, et al. Sphingolipid metabolism regulates development and lifespan in Caenorhabditis elegans[J]. Mechanisms of ageing and development, 2014, 143/144: 9-18.

[308] JUN H J, KIM J, HOANG M H, et al. Hepatic lipid accumulation alters global histone h3 lysine 9 and 4 trimethylation in the peroxisome proliferator-activated receptor alpha network[J]. PLoS one, 2012, 7(9): 1-8.

[309] MENUZ V, HOWELL K S, GENTINA S, et al. Protection of C. elegans

from anoxia by HYL-2 ceramide synthase[J]. Science, 2009, 324(5925): 381-384.

[310] MOSBECH M B, KRUSE R, HARVALD E B, et al. Functional loss of two ceramide synthases elicits autophagy-dependent lifespan extension in C. elegans[J]. PLoS one, 2017, 8(7): e70087.

[311] DENG X, YIN X, ALLAN R, et al. Ceramide biogenesis is required for radiation-induced apoptosis in the germ line of C. elegans[J]. Science, 2008, 322(5898): 110-115.

[312] WALKER A K, JACOBS R L, WATTS J L, et al. A conserved SREBP-1/phosphatidylcholine feedback circuit regulates lipogenesis in metazoans[J]. Cell, 2011, 147(4): 840-852.

[313] FU S, YANG L, LI P, et al. Aberrant lipid metabolism disrupts calcium homeostasis causing liver endoplasmic reticulum stress in obesity[J]. Nature, 2011, 473(7348): 528-531.

[314] ZOU C G, TU Q, NIU J, et al. The DAF-16/FOXO transcription factor functions as a regulator of epidermal innate immunity[J]. PLoS pathogens, 2013, 9(10): e1003660.

[315] HENDERSON S T, JOHNSON T E. daf-16 integrates developmental and environmental inputs to mediate aging in the nematode Caenorhabditis elegans[J]. Current biology, 2001, 11(24): 1975-1980.

[316] VECCHI C, MONTOSI G, ZHANG K, et al. ER stress controls iron metabolism through induction of hepcidin[J]. Science, 2009, 325(5942): 877-880.

[317] HOTAMISLIGIL G S. Endoplasmic reticulum stress and the inflammatory basis of metabolic disease[J]. Cell, 2010, 140(6): 900-917.

附 录

F.1 QPCR引物

引物	序列
snb-1 Forward	GCAAGTATTGGTGGAAGA
snb-1 Reverse	ACGATGATGATAATAAGAATGAC
let-607 Forward	CGGATCTGATTGGAATGGAC
let-607 Reverse	GATGAGGGAGATGTGCTTG
T24B8.5 Forward	TTGTGATTGTGCTTGTAG
T24B8.5 Reverse	CAACCACTTCTAACATCTG
hsp-3 Forward	TCATGGGCATGCGAATCAAC
hsp-3 Reverse	AGTGCCGAACACCATCGTAA
hsp-4 Forward	GAAACAGAATCACTCCATCAT
hsp-4 Reverse	CAGTGCTTGATGTCTTGTT
pek-1 Forward	ATGGAGGATCTGACAGAACT
pek-1 Reverse	CTCAATTCCTCCTGATGAAGAG
atf-4 Forward	CCATTCCACCCCACAATAT
atf-4 Reverse	GGAAGTTGACATCGGAGTT
eif-2α Forward	CAACATTCAATCAGGAGAGTCT
eif-2α Reverse	CGTTTCGTGGACCTTCAAT
ire-1 Forward	TGGAAACTCTATCATCAGCGT
ire-1 Reverse	CCACGTATTCACTTCAGGC
xbp-1s Forward	TGCCTTTGAATCAGCAGTGG
xbp-1s Reverse	ACCGTCTGCTCCTTCCTCAATG
atf-6 Forward	GAATCACAAGAATCGACCTCT
atf-6 Reverse	CCCACATTTCCTGGTCAT
pdi-1 Forward	TGAAGATGGAGTCGCTCTTAT

引物	序列
pdi-1 Reverse	AGGTCTCCTCCGACGAT
pdi-2 Forward	GACACCACCTCCGATGAT
pdi-2 Reverse	TTGGGTGAGCTTCTCGT
daf-16 Forward	AAGCCGATTAAGACGGAACC
daf-16 Reverse	GTAGTGGCATTGGCTTGAAG
sod-3 Forward	CCAACCAGCGCTGAAATTCAATGG
sod-3 Reverse	GGAACCGAAGTCGCGCTTAATAGT
aha-1 Forward	TTCCATGTACTGTGTCTGC
aha-1 Reverse	TTCACTGGCTTGAGGTTG
gsk-3 Forward	ACGCATTCTTTGATGAGCT
gsk-3 Reverse	GCTTGTCGTTGAAACTTCAC
sms-5 Forward	AGAACCTGCTTCACGCCT
sms-5 Reverse	GACGCCCACCATAGCAGTAG
mtl-1 Forward	ATGGCTTGCAAGTGTGACTG
mtl-1 Reverse	CACATTTGTCTCCGCACTTG
$sams$-1 Forward	CGGATATGCAACCGACGA
$sams$-1 Reverse	GACCACAACAGTGTGAACG
ckb-1 Forward	TGCCGTGTACCCCAAAT
ckb-1 Reverse	TTCATAGTCGCCACCGC
$pcyt$-1 Forward	TCATGGGCATGCGAATCAAC
$pcyt$-1 Reverse	AGTGCCCAACACCATCGTAA
$cept$-1 Forward	GCACCGTCATGGGCTTAT
$cept$-1 Reverse	CTTGCGTCATGCTATCACAAC
pmt-2 Forward	ATAAGGTGACCGAGGGAC
pmt-2 Reverse	TCGGCGTTGCGAATAGT
itr-1 Forward	TTGGAACACTGGTGGCTA
itr-1 Reverse	GGAAGCTGCACAGATTTCAT
egl-8 Forward	GAAAGTACGTGTCTGGAACT
egl-8 Reverse	GCCATGCGTGATCCATAT
sca-1 Forward	GTCGGAGTCGTCGGAATG
sca-1 Reverse	GGCAGGCTTCAGATTGTTG
pkc-2 Forward	GACCGTTTGTATTTCGTGAT
pkc-2 Reverse	TTTATATGCCCGTCTCGTT

F.2 ChIP-seq结果

acbp-6，*acdh*-1，*acin*-1，*acs*-13，*acs*-14，*acs*-4，*acs*-6，*act*-5，*adss*-1，*agt*-2，*ain*-1，*aka*-1，*alh*-12，*anr*-17，*anr*-45，*anr*-53，*anr*-54，*anr*-58，*ant*-1.1，*apb*-1，*aqp*-10，*arf*-1.2，*arf*-3，*ari*-1.4，*arl*-7，*ash*-2，*asp*-1，*atad*-3，*atfs*-1，*atgl*-1，*atx*-2，*B0035.22*，*B0250.5*，*B0261.1*，*B0303.7*，*B0350.t1*，*B0365.10*，*B0393.12*，*B0393.3*，*B0403.6*，*B0416.5*，*B0491.5*，*B0507.3*，*B0507.3*，*B0511.11*，*B0511.2*，*B0524.10*，*B0546.4*，*bag*-1，*bet*-2，*bris*-1，*btf*-1，*C01G10.7*，*C01G12.12*，*C01G6.5*，*C02B4.8*，*C02F12.t1*，*C03A3.t1*，*C03H12.1*，*C04A11.t1*，*C04E7.t1*，*C05D2.15*，*C05G5.2*，*C06A5.3*，*C06G1.t2*，*C07A12.19*，*C07G2.4*，*C07H6.2*，*C08B11.13*，*C08D8.3*，*C08E8.12*，*C08F8.13*，*C09B8.4*，*C09G1.2*，*C09G9.1*，*C10F3.15*，*C10F3.16*，*C10G8.9*，*C12C8.8*，*C13G3.10*，*C14A11.9*，*C14B1.3*，*C14B9.2*，*C14F11.14*，*C15C8.7*，*C16C2.7*，*C16C8.t2*，*C17C3.23*，*C17H11.10*，*C17H12.35*，*C17H12.36*，*C18A11.13*，*C18B12.6*，*C18B2.8*，*C18C4.13*，*C18C4.17*，*C18E9.2*，*C24F3.2*，*C24G6.10*，*C25F6.t3*，*C26B2.12*，*C26B9.1*，*C26C6.12*，*C26G2.t1*，*C27C12.4*，*C27D6.14*，*C27D6.14*，*C27H6.5*，*C28G1.t1*，*C28H8.14*，*C29F7.3*，*C30C11.14*，*C30E1.9*，*C32D5.14*，*C33D9.3*，*C33G3.t2*，*C34B4.2*，*C34C12.2*，*C34D10.4*，*C34E7.6*，*C35B1.2*，*C36A4.12*，*C36E8.10*，*C36E8.11*，*C37C3.16*，*C39F7.7*，*C40H1.11*，*C40H5.t1*，*C41A3.6*，*C41C4.11*，*C42D4.17*，*C42D8.9*，*C44C10.11*，*C44C11.5*，*C46H11.3*，*C47E8.4*，*C48B4.13*，*C49A9.10*，*C49F8.10*，*C49F8.7*，*C50H2.13*，*C51F7.4*，*C52B11.5*，*C52E12.1*，*C53B7.8*，*C53C7.6*，*C53C7.t1*，*C53D5.5*，*C53H9.4*，*C56C10.15*，*cap*-2，*cars*-1，*cct*-4，*cct*-4，*ccz*-1，*cdc*-48.1，*cdk*-5，*cdk*-7，*cdkr*-3，*cec*-1，*ced*-5，*cel*-1，*cept*-2，*cgh*-1，*cgp*-1，*cids*-2，*clec*-150，*clec*-9，*clic*-1，*cnk*-1，*cnt*-2，*cogc*-2，*copa*-1，*copd*-1，*cope*-1，*COX2*，*cox*-5A，*cox*-5B，*cpn*-3，*cpr*-4，*cpt*-2，*cpz*-1，*cTel17.1*，*cTel52S.1*，*cTel79B.1*，*cTel7X.2*，*cts*-1，*cyn*-10，*cyn*-4，*D1007.18*，*D1046.15*，*D1046.6*，*D1054.18*，*D2023.4*，*D2045.2*，*daf*-14，*daf*-5，*dbt*-1，*dcs*-1，*dct*-11，*dct*-16，*dct*-18，*ddl*-3，*ddl*-3，*ddp*-1，*dgat*-2，*dhhc*-10，*dhs*-25，*DL2.2*，*dnc*-6，*dnj*-12，*dnj*-20，*dnj*-24，*dnj*-28，*dnj*-29，*dnj*-7，*dnj*-9，*dnsn*-1，*dpm*-3，*dpy*-11，*dpyd*-1，*drr*-2，*dyn*-1，*E03E2.t1*，*E03E2.t3*，*ears*-2，*eas*-1， *ech*-1.2, *eef*-1B.2, *eef*-1G, *eef*-2, *eel*-1, *egl*-30, *egl*-4, *egl*-44, *egl*-44, *egl*-9, *eif*-3.B, *eif*-3.C, *eif*-3.E, *ekl*-4, *ekl*-4, *elk*-2, *emo*-1, *enpl*-1, *epg*-7, *epg*-9, *erd*-2.1, *erm*-1, *erp*-44.1, *erv*-46, *ets*-4, *exoc*-7, *F01D5.12*, *F01F1.3*, *F01G12.t1*, *F01G4.5*, *F02D10.11*, *F02E9.5*, *F07F6.4*, *F07H5.13*, *F07H5.5*, *F08F1.4*, *F08G2.13*, *F09A5.11*, *F09B9.8*, *F09B9.t1*, *F09C8.3*, *F09E5.10*, *F09E5.22*, *F09E5.9*, *F09E5.9*, *F09F7.6*, *F09G2.14*, *F10E9.5*, *F10G7.9*, *F11A10.9*, *F11C7.8*, *F12F6.11*, *F12F6.14*, *F13A7.14*, *F13B9.1*, *F13E6.1*, *F13G3.10*, *F14E5.2*, *F15B9.15*, *F15G9.8*, *F16G10.6*, *F17C11.20*, *F17C11.24*, *F17E5.2*, *F17H10.1*, *F17H10.t1*, *F18A11.3*, *F18F11.1*, *F18G5.6*, *F18H3.4*, *F18H3.4*, *F19B2.5*, *F19H6.6*, *F20B6.4*, *F20B6.7*, *F20D1.13*, *F20D12.12*, *F21D5.7*, *F22A3.15*, *F22B7.14*, *F22E10.6*, *F22F1.t3*, *F22H10.2*, *F23C8.5*, *F25B5.5*, *F25D1.11*, *F25D7.11*, *F25G6.9*, *F25H2.15*, *F25H9.7*, *F26D10.23*, *F26D10.24*, *F26F12.11*, *F26F2.7*, *F26F4.14*, *F26H9.5*, *F27C1.17*, *F27D4.3*, *F27D4.9*, *F28A12.6*, *F29B9.11*, *F29G6.4*, *F30H5.4*, *F31F7.1*, *F32B5.6*, *F32B5.9*, *F32D8.14*, *F32E10.10*, *F33C8.9*, *F33D11.t2*, *F35D11.4*, *F36A2.7*, *F37F2.2*, *F37H8.2*, *F38A1.8*, *F38E11.10*, *F38E11.25*, *F38G1.t2*, *F39B2.14*, *F40E10.6*, *F40E3.3*, *F40F11.6*, *F40F9.14*, *F42A10.9*, *F42D1.11*, *F42D1.6*, *F42D1.7*, *F42D1.t1*, *F42G4.6*, *F43C9.1*, *F44B9.11*, *F44E5.14*, *F44E5.4*, *F44E7.17*, *F44E7.4*, *F44E7.9*, *F45H7.t1*, *F46B6.15*, *F46G10.1*, *F46G11.t2*, *F46H5.7*, *F47B7.2*, *F47G4.14*, *F47G4.9*, *F47G6.t1*, *F47G9.1*, *F48A9.6*, *F48C1.12*, *F48C1.13*, *F49E7.5*, *F52C9.t1*, *F52D10.7*, *F52D10.9*, *F52E10.t5*, *F52H2.7*, *F52H2.t1*, *F53A3.8*, *F53A3.8*, *F53A9.9*, *F53E10.8*, *F53H4.4*, *F54C9.17*, *F54C9.3*, *F54D5.22*, *F54D8.7*, *F54F12.6*, *F54F2.6*, *F55A11.1*, *F55A11.10*, *F55A11.6*, *F55F3.t1*, *F55G7.5*, *F55G7.t2*, *F56A8.3*, *F56A8.9*, *F56C4.4*, *F56C9.10*, *F56F3.11*, *F56G4.6*, *F57B10.16*, *F57B10.17*, *F57F5.7*, *F57H12.10*, *F58A3.6*, *F58D12.2*, *F58D5.5*, *F58D5.t1*, *F59A2.5*, *F59A7.8*, *F59B1.2*, *F59C12.t3*, *F59C6.15*, *F59F4.2*, *fat*-2, *fat*-3, *fat*-6, *fbp*-1, *fbxa*-190, *fbxa*-55, *fbxa*-75, *fbxb*-58, *fkb*-1, *fkb*-5, *frm*-10, *ftt*-2, *fust*-1, *gakh*-1, *gdi*-1, *gei*-4, *gfm*-1, *glr*-7, *gob*-1, *gsa*-1, *gsp*-2, *gst*-9, *gsy*-1, *gtr*-1, *guk*-1, *H06A10.t1*, *H08J11.t3*, *H09F14.1*, *H09I01.2*, *H13N06.9*, *H15M21.1*, *H15M21.1*, *H37A05.4*, *haf*-3, *hal*-3, *hat*-1, *hcp*-3, *hcp*-6, *hda*-4, *hex*-3, *hif*-1, *his*-24, *his*-37, *hphd*-1, *hphd*-1, *hpo*-34, *hpr*-9, *hrpk*−1，*hrpu*−2，*hsp*−3，*hsp*−4，*hsr*−9，*hst*−2，*hxk*−1，*icd*−2，*idh*−1，*ife*−2，*iff*−2，*ikke*−1，*ile*−1，*ilys*−5，*imb*−3，*imb*−3，*immp*−1，*ins*−39，*ipgm*−1，*ipla*−6，*ivd*−1，*K02A4.t*1，*K02A6.3*，*K02B12.7*，*K02D3.4*，*K02D7.1*，*K02F3.13*，*K02G10.9*，*K03A1.4*，*K03C7.11*，*K03C7.14*，*K03H4.3*，*K04F10.7*，*K04G7.1*，*K05C4.2*，*K07A1.19*，*K07C5.13*，*K07C5.6*，*K07D4.1*，*K07D4.1*，*K07H8.16*，*K08B5.2*，*K08D10.15*，*K08D8.1*，*K08F11.1*，*K08F11.6*，*K08F11.7*，*K09D9.1*，*K10D2.5*，*K11D12.14*，*K11D9.10*，*K11D9.8*，*K12H4.5*，*kbp*−4，*kin*−10，*kin*−14，*klp*−13，*lars*−1，*lbp*−5，*lbp*−9，*lbp*−9，*let*−607，*lgc*−26，*lgg*−2，*lgg*−2，*lin*−35，*lin*−53，*lin*−59，*lin*−66，*lin*−9，*linc*−100，*linc*−103，*linc*−105，*linc*−128，*linc*−135，*linc*−44，*linc*−78，*linc*−81，*linc*−90，*linc*−91，*lpd*−9，*lron*−13，*lron*−7，*ltah*−1.2，*ltah*−1.2，*M01H9.6*，*M02B1.2*，*M02B1.3*，*M02B1.3*，*M02F4.t*1，*M03F4.13*，*M04B2.10*，*M04F3.7*，*M106.2*，*M106.3*，*M106.6*，*M117.12*，*M162.13*，*M163.1*，*M163.13*，*M6.11*，*M7.15*，*mak*−1，*malt*−1，*manf*−1，*mans*−3，*mboa*−2，*mdh*−2，*mdt*−26，*mec*−6，*memb*−1，*met*−1，*mev*−1，*mig*−5，*mir*−2214，*mir*−2221，*mir*−229，*mir*−239.2，*mir*−34，*mir*−4811，*mir*−4936，*mir*−53，*mir*−5547，*mir*−5549，*mir*−5595，*mir*−800，*mir*−8186.2，*mml*−1，*mog*−4，*mom*−1，*mpz*−5，*mrpl*−51，*mrpl*−53，*mrpl*−55，*mrpr*−1，*mtcu*−2，*mtcu*−2，*mxl*−3，*mxl*−3，*nbet*−1，*nck*−1，*ncl*−1，*ncs*−2，*ndk*−1，*nduf*−7，*nhr*−14，*nhr*−19，*nhr*−20，*nhr*−35，*nhr*−49，*nlp*−40，*nlp*−55，*nlp*−55，*nmad*−1，*nol*−10，*nol*−56，*nol*−58，*npp*−18，*npp*−20，*npp*−3，*npp*−6，*npp*−8，*nspc*−17，*ntl*−11，*ntl*−2，*nuc*−1，*nucb*−1，*num*−1，*ogt*−1，*ostb*−1，*ostd*−1，*otub*−2，*pam*−1，*pamn*−1，*pcn*−1，*pcp*−1，*pde*−6，*pdhk*−2，*pdi*−1，*pdi*−6，*pek*−1，*pes*−10，*pes*−10，*pgp*−2，*plrg*−1，*pms*−2，*pmt*−1，*pmt*−1，*pno*−1，*pno*−1，*pqm*−1，*pqn*−39，*pqn*−63，*pqn*−68，*pqn*−70，*prmt*−5，*prp*−31，*R01H5.t*1，*R03E1.t*2，*R03E9.10*，*R03E9.6*，*R03E9.7*，*R03E9.8*，*R03E9.8*，*R03G5.8*，*R04A9.t*1，*R04D3.t*2，*R04E5.9*，*R05D7.9*，*R05G6.5*，*R06B9.5*，*R106.6*，*R107.5*，*R10E11.10*，*R10E11.5*，*R11A5.6*，*R12B2.9*，*R12C12.6*，*R12C12.8*，*R12E2.13*，*R12E2.18*，*R13A5.11*，*R148.3*，*R166.2*，*R186.8*，*rad*−50，*rbg*−1，*rde*−12，*rer*−1，*ret*−1，*rga*−1，*rheb*−1，*rib*−2，*riok*−1，*rme*−4，*rncs*−1，*rncs*−1，*rpa*−2，*rpac*−19，*rpb*−11，*rpb*−8，*rpl*−12，*rpl*−14，*rpl*−15，*rpl*−22，*rpl*−24.2，*rpl*−29，*rpl*−3，*rpl*−35，*rpl*−41.2，*rpl*−6，*rpl*−7A，*rpn*−11，*rpr*−1，*rps*−14，*rps*−15，*rps*−24，*rps*−28，*rps*−4，*rrn*−2.1，*rrn*−4.15，*rrn*−4.2，*rsd*−3，*rsp*−

1, *rsp-6*, *rte-1*, *ruvb-2*, *sar-1*, *sco-1*, *sdc-2*, *sec-12*, *sec-61*, *sek-3*, *sel-9*, *set-2*, *set-25*, *set-4*, *sft-4*, *ska-3*, *slc-25A21*, *slrp-1*, *slrp-1*, *sls-1.12*, *sls-1.4*, *sls-1.6*, *sls-2.1*, *sls-2.8*, *sly-1*, *smu-1*, *smy-2*, *snrp-40.1*, *sodh-1*, *sor-1*, *spat-1*, *spc-1*, *spcs-2*, *spe-48*, *sqd-1*, *srg-64*, *srh-2*, *srh-70*, *sri-5*, *srpa-68*, *srpa-72*, *srpr-1.1*, *srv-1*, *srw-45*, *srw-78*, *srz-98*, *stau-1*, *stc-1*, *subs-4*, *sucl-1*, *sup-6*, *sup-7*, *suro-1*, *sym-4*, *T01C1.t1*, *T02E1.9*, *T02G6.10*, *T03E6.t1*, *T03F7.8*, *T03G6.t1*, *T04C10.11*, *T04C10.8*, *T04C12.8*, *T04C4.3*, *T04F8.2*, *T05B11.1*, *T05B11.10*, *T05H4.7*, *T06C12.16*, *T06D8.11*, *T07E3.4*, *T07F10.3*, *T07H6.t2*, *T08G2.9*, *T09F3.6*, *T09F3.6*, *T10B10.13*, *T10B10.3*, *T10B5.3*, *T10B9.11*, *T12D8.5*, *T12G3.9*, *T14G10.5*, *T14G8.3*, *T14G8.5*, *T14G8.5*, *T15B12.t1*, *T15H9.8*, *T15H9.8*, *T16G1.18*, *T18D3.t1*, *T19B4.3*, *T19B4.8*, *T19C11.3*, *T19C3.11*, *T20D3.13*, *T21C12.9*, *T21D12.7*, *T22A3.t1*, *T23B5.3*, *T23F1.t1*, *T23F11.10*, *T24B8.22*, *T24D8.6*, *T25B6.t2*, *T25C8.1*, *T25E4.3*, *T26A5.10*, *T26C5.5*, *T26H2.15*, *T27A3.12*, *T27A8.8*, *T27E4.13*, *T28B4.t1*, *T28B8.1*, *T28C6.7*, *T28F4.5*, *T28F4.7*, *T28F4.8*, *T28H10.2*, *tag-124*, *tag-260*, *tasp-1*, *tba-2*, *tbc-3*, *tcc-1*, *tdpt-1*, *toc-1*, *tomm-20*, *tos-1*, *tos-1*, *tps-2*, *tpst-1*, *trap-1*, *trap-2*, *trap-4*, *ttr-37*, *ttr-45*, *ttr-53*, *uba-5*, *ubc-3*, *ubc-8*, *ubl-5*, *ucr-2.2*, *ufbp-1*, *ugt-1*, *ugt-49*, *ugt-50*, *ugt-6*, *unc-23*, *unc-69*, *unc-78*, *unc-85*, *upp-1*, *uso-1*, *usp-14*, *vha-1*, *vha-12*, *vha-13*, *vha-19*, *vit-2*, *vit-3*, *vit-5*, *vpr-1*, *vps-25*, *vps-39*, *vps-4*, *vps-60*, *W01C9.6*, *W01D2.8*, *W01H2.8*, *W02D3.13*, *W02D7.6*, *W02H3.t1*, *W02H3.t1*, *W03C9.5*, *W03C9.9*, *W03F8.4*, *W03G11.t1*, *W04A4.5*, *W04C9.4*, *W04D12.2*, *W05B2.8*, *W05H9.1*, *W06A7.6*, *W06B11.7*, *W06F12.3*, *W06H8.14*, *W06H8.15*, *W08E3.5*, *W08G11.6*, *W09D10.1*, *W09D6.5*, *W09G10.t3*, *W09G3.8*, *W09H1.7*, *wapl-1*, *wdr-23*, *wip-1*, *wve-1*, *xbp-1*, *xnp-1*, *xpd-1*, *Y102A11A.9*, *Y102F5A.1*, *Y105C5A.1268*, *Y105E8A.3*, *Y106G6A.1*, *Y111B2A.12*, *Y111B2A.34*, *Y116A8A.10*, *Y119C1B.3*, *Y119C1B.3*, *Y119D3B.10*, *Y17G7B.10*, *Y17G7B.17*, *Y17G7B.8*, *Y17G9B.5*, *Y18D10A.4*, *Y19D10B.4*, *Y20F4.4*, *Y22D7AL.10*, *Y23H5B.13*, *Y25C1A.7*, *Y25C1A.8*, *Y26E6A.3*, *Y32B12B.2*, *Y32F6A.6*, *Y32H12A.9*, *Y34F4.1*, *Y37A1B.331*, *Y38A10A.10*, *Y38C1AA.12*, *Y38E10A.t1*, *Y38F2AR.14*, *Y38F2AR.16*, *Y39B6A.43*, *Y39B6A.5*, *Y39E4B.6*, *Y39E4B.6*, *Y39G10AR.24*, *Y41C4A.32*, *Y41D4A.4*, *Y41D4A.5*, *Y42H9AR.1*,

*Y43D4A.t*1, *Y43F8B.*2, *Y43F8B.*2, *Y43F8B.*26, *Y43F8C.*22, *Y47C4A.t*2, *Y48A6B.*7, *Y48B6A.*1, *Y48B6A.*18, *Y48C3A.*3, *Y48G10A.*1, *Y48G1C.*13, *Y48G1C.*8, *Y48G8AL.*13, *Y49A3A.*3, *Y49G5A.*1, *Y50D4B.*2, *Y50E8A.*8, *Y51A2D.*13, *Y51A2D.*32, *Y51F10.*7, *Y51H4A.*15, *Y52B11C.*1, *Y53C12A.*11, *Y53F4B.*47, *Y53F4B.*48, *Y54E10BR.*2, *Y54E10BR.*2, *Y54F10AM.*15, *Y54G11A.*2, *Y54G2A.*18, *Y54G2A.*28, *Y54G2A.*48, *Y54G2A.*70, *Y55F3AR.*2, *Y55F3BR.*11, *Y56A3A.*18, *Y56A3A.*22, *Y56A3A.*7, *Y57A10A.*14, *Y57E12AL.*6, *Y57G11C.*1141, *Y57G11C.*9, *Y59A8B.*24, *Y5F2A.*3, *Y60A3A.*29, *Y62E10A.*13, *Y62E10A.*3, *Y62F5A.*12, *Y62H9A.*7, *Y64H9A.*1, *Y65A5A.*2, *Y65B4BR.*10, *Y67D8A.*9, *Y69E1A.*13, *Y6B3B.*4, *Y70D2A.*5, *Y70D2A.*6, *Y70G10A.*1, *Y71F9AL.*19, *Y71F9AM.*7, *Y71G12A.*5, *Y71H10B.*1, *Y71H2AM.*11, *Y71H2AM.*20, *Y73B6BL.*29, *Y73F4A.*1, *Y73F4A.*1, *Y73F8A.*1173, *Y73F8A.*27, *Y74C9A.*6, *Y74C9A.*6, *Y74C9A.*9, *Y75B12B.*12, *Y75B8A.*63, *Y75D11A.*6, *Y76A2B.*5, *Y77E11A.*2, *Y7A5A.*11, *Y7A5A.*3, *Y7A5A.t*4, *Y81B9A.*1, *Y82E9BR.*27, *Y82E9BR.*3, *Y95B8A.*16, *Y95B8A.*2, *Y95B8A.*6, *Y95B8A.*8, *ZC204.*13, *ZC262.*13, *ZC412.*15, *ZC412.t*2, *ZC449.t*1, *ZC506.*1, *ZC53.*1, *zfp*-3, *zip*-10, *zip*-11, *zip*-11, *zipt*-9, *ZK1098.*1, *ZK1098.*2, *ZK112.*14, *ZK112.*4, *ZK1193.*10, *ZK1290.*22, *ZK177.*8, *ZK185.*5, *ZK20.*8, *ZK265.*10, *ZK265.*11, *ZK484.*9, *ZK546.*20, *ZK550.*3, *ZK593.*9, *ZK632.*10, *ZK637.*2, *ZK643.*9, *ZK673.*2, *ZK757.*7, *ZK829.*13, *ZK836.*7, *ZK899.*10, *ZK899.t*1, *ZK994.*7, *ztf*-13, *zyg*-11, 1-*Jun*, 21*ur*-10291, 21*ur*-10977, 21*ur*-11069, 21*ur*-11151, 21*ur*-11631, 21*ur*-11746, 21*ur*-11746, 21*ur*-11746, 21*ur*-12080, 21*ur*-12279, 21*ur*-12394, 21*ur*-12946, 21*ur*-12994, 21*ur*-13313, 21*ur*-13313, 21*ur*-13424, 21*ur*-13751, 21*ur*-14797, 21*ur*-14870, 21*ur*-15151, 21*ur*-15151, 21*ur*-15257, 21*ur*-15362, 21*ur*-3629, 21*ur*-6618, 21*ur*-8200, 21*ur*-8451, 21*ur*-865, 2*RSSE.*7, 3*R*5。

F.3 病原菌感染实验统计结果

重复1[①]为文中用图数据,其余为生物学重复。

图序	线虫品系 及处理方法	平均寿命 ± SEM/天	非正常 死亡数/总数	P值
图3.4 重复1	control RNAi	4.00 ± 0.18	0/35	
	let-607 RNAi	4.97 ± 0.17	0/35	<0.001[a]
图3.4 重复2	control RNAi	4.00 ± 0.18	0/35	
	let-607 RNAi	4.97 ± 0.16	0/35	<0.001[a]
图3.4 重复3	control RNAi	4.09 ± 0.15	0/35	
	let-607 RNAi	4.94 ± 0.17	0/35	<0.001[a]
图3.4 重复4	control RNAi	3.75 ± 0.13	0/36	
	let-607 RNAi	4.46 ± 0.15	0/37	<0.001[a]
图3.4 重复5	control RNAi	3.86 ± 0.11	0/36	
	let-607 RNAi	4.49 ± 0.13	0/37	<0.001[a]
图3.8(a) 重复1	WT, control RNAi	3.67 ± 0.12	0/36	
	WT, *let*-607 RNAi	4.44 ± 0.13	0/36	<0.001[a]
	daf-16, control RNAi	2.89 ± 0.12	0/35	
	daf-16, *let*-607 RNAi	3.11 ± 0.14	0/35	0.213[a]
图3.8(a) 重复2	WT, control RNAi	3.49 ± 0.12	0/35	
	WT, *let*-607 RNAi	4.50 ± 0.13	0/36	<0.001[a]
	daf-16, control RNAi	3.00 ± 0.11	0/35	
	daf-16, *let*-607 RNAi	3.06 ± 0.13	0/35	0.605[a]
图3.8(a) 重复3	WT, control RNAi	3.53 ± 0.11	0/36	
	WT, *let*-607 RNAi	4.23 ± 0.16	0/35	<0.001[a]
	daf-16, control RNAi	3.00 ± 0.13	0/35	
	daf-16, *let*-607 RNAi	3.06 ± 0.08	0/35	0.814[a]

---

① 生物学实验,同一实验内容需要做三次生物学重复,每次重复都有一个实验结果图。本书中所展示的图片均来自原始数据的第一组图,其余两组重复不做图,但需要展示原始数据,证明有三次生物学重复实验。

图序	线虫品系及处理方法	平均寿命 ± SEM/天	非正常死亡数/总数	P 值
图3.8（a）重复4	WT, control RNAi	4.19 ± 0.16	0/36	
	WT, let-607 RNAi	5.11 ± 0.19	0/35	<0.001[a]
	daf-16, control RNAi	3.51 ± 0.11	0/37	
	daf-16, let-607 RNAi	3.63 ± 0.13	0/35	0.605[a]
图3.8（a）重复5	WT, control RNAi	3.94 ± 0.17	0/35	
	WT, let-607 RNAi	5.06 ± 0.18	0/36	<0.001[a]
	daf-16, control RNAi	3.44 ± 0.12	0/36	
	daf-16, let-607 RNAi	3.57 ± 0.14	0/35	0.814[a]
图3.8（b）重复1	WT, control RNAi	3.94 ± 0.17	0/35	
	WT, let-607 RNAi	5.06 ± 0.18	0/36	<0.001[a]
	hsf-1, control RNAi	2.87 ± 0.12	0/39	
	hsf-1, let-607 RNAi	3.57 ± 0.14	0/37	<0.001[a]
图3.8（b）重复2	WT, control RNAi	4.19 ± 0.16	0/36	
	WT, let-607 RNAi	5.11 ± 0.19	0/35	<0.001[a]
	hsf-1, control RNAi	2.86 ± 0.12	0/37	
	hsf-1, let-607 RNAi	3.51 ± 0.14	0/35	<0.001[a]
图3.8（b）重复3	WT, control RNAi	3.86 ± 0.15	0/36	
	WT, let-607 RNAi	5.16 ± 0.17	0/37	<0.001[a]
	hsf-1, control RNAi	2.86 ± 0.11	0/36	
	hsf-1, let-607 RNAi	3.49 ± 0.11	0/35	<0.001[a]
图3.8（b）重复4	WT, control RNAi	3.49 ± 0.12	0/35	
	WT, let-607 RNAi	4.50 ± 0.13	0/36	<0.001[a]
	hsf-1, control RNAi	2.47 ± 0.10	0/36	
	hsf-1, let-607 RNAi	3.03 ± 0.12	0/35	<0.001[a]
图3.8（b）重复5	WT, control RNAi	3.53 ± 0.11	0/36	
	WT, let-607 RNAi	4.23 ± 0.16	0/35	<0.001[a]
	hsf-1, control RNAi	2.40 ± 0.10	0/35	
	hsf-1, let-607 RNAi	3.11 ± 0.14	0/35	<0.001[a]

注：[a]表示该组 P 值为相同处理组内或相同线虫品系间比较。

F.4 活性氧应激实验统计结果

重复1为文中用图数据，其余为生物学重复。

图序	线虫品系及处理方法	平均寿命 ± SEM/天	非正常死亡数/总数	P值
图3.9（a）重复1	control RNAi	10.00 ± 0.33	0/31	
	let-607 RNAi	13.60 ± 0.19	0/35	<0.001[a]
图3.9（a）重复2	control RNAi	8.93 ± 0.23	0/30	
	let-607 RNAi	13.14 ± 0.25	0/35	<0.001[a]
图3.9（a）重复3	control RNAi	9.14 ± 0.25	0/37	
	let-607 RNAi	13.53 ± 0.41	0/34	<0.001[a]
图3.10 重复1	VP303, control RNAi	6.14 ± 0.06	0/73	
	VP303, let-607 RNAi	7.37 ± 0.15	0/63	<0.001[a]
图3.10 重复2	VP303, control RNAi	6.03 ± 0.03	2/62	
	VP303, let-607 RNAi	6.73 ± 0.16	0/49	<0.001[a]
图3.10 重复3	VP303, control RNAi	6.27 ± 0.09	0/52	
	VP303, let-607 RNAi	7.80 ± 0.16	0/59	<0.001[a]
图3.10 重复1	DCL569, control RNAi	7.27 ± 0.22	0/33	
	DCL569, let-607 RNAi	6.79 ± 0.17	0/33	0.088[a]
图3.10 重复2	DCL569, control RNAi	7.51 ± 0.14	0/37	
	DCL569, let-607 RNAi	7.80 ± 0.09	0/49	0.081[a]
图3.10 重复3	DCL569, control RNAi	6.36 ± 0.13	0/33	
	DCL569, let-607 RNAi	6.21 ± 0.10	0/38	0.358[a]
图3.10 重复1	WM118, control RNAi	7.41 ± 0.36	0/37	
	WM118, let-607 RNAi	7.09 ± 0.31	0/35	0.413[a]
图3.10 重复2	WM118, control RNAi	6.00 ± 0.00	0/43	
	WM118, let-607 RNAi	6.05 ± 0.05	0/37	0.281[a]
图3.10 重复3	WM118, control RNAi	6.28 ± 0.12	0/36	
	WM118, let-607 RNAi	6.11 ± 0.07	0/37	0.222[a]
图3.10 重复1	NR222, control RNAi	6.92 ± 0.27	0/37	
	NR222, let-607 RNAi	7.08 ± 0.32	0/39	0.631[a]

图序	线虫品系及处理方法	平均寿命 ± SEM/天	非正常死亡数/总数	P值
图3.10 重复2	NR222, control RNAi	7.79 ± 0.31	0/38	
	NR222, let-607 RNAi	7.79 ± 0.31	0/39	0.970[a]
图3.10 重复3	NR222, control RNAi	7.49 ± 0.20	0/35	
	NR222, let-607 RNAi	7.31 ± 0.23	0/35	0.626[a]
图3.26(a) 重复1	control RNAi	7.68 ± 0.20	5/49	
	let-607 RNAi	10.32 ± 0.27	0/50	<0.001[a]
图3.26(a) 重复2	control RNAi	8.62 ± 0.18	0/39	
	let-607 RNAi	10.50 ± 0.40	1/37	<0.001[a]
图3.26(a) 重复3	control RNAi	9.43 ± 0.18	0/42	
	let-607 RNAi	11.84 ± 0.34	0/50	<0.001[a]
图3.26(a) 重复4	control RNAi	8.33 ± 0.21	0/54	
	let-607 RNAi	10.83 ± 0.30	3/54	<0.001[a]
图3.26(a) 重复1	WT, control RNAi	8.93 ± 0.23	0/30	
	WT, let-607 RNAi	13.14 ± 0.25	0/35	<0.001[a]
	daf-16, control RNAi	8.81 ± 0.21	0/32	
	daf-16, let-607 RNAi	10.74 ± 0.26	0/35	<0.001[a], <0.001[b]
图3.26(a) 重复2	WT, control RNAi	9.14 ± 0.25	0/37	
	WT, let-607 RNAi	13.53 ± 0.41	0/34	<0.001[a]
	daf-16, control RNAi	8.68 ± 0.24	0/41	
	daf-16, let-607 RNAi	10.34 ± 0.58	0/35	<0.001[a], <0.001[b]
图3.26(a) 重复3	WT, control RNAi	10.00 ± 0.33	0/31	
	WT, let-607 RNAi	13.60 ± 0.19	0/35	<0.001[a]
	daf-16, control RNAi	9.88 ± 0.26	0/33	
	daf-16, let-607 RNAi	12.29 ± 0.35	0/35	<0.001[a], <0.001[b]
图3.26(d) 重复1	WT, control RNAi	8.11 ± 0.25	0/35	
	WT, let-607 RNAi	12.69 ± 0.40	2/35	<0.001[a]
	hsf-1, control RNAi	6.86 ± 0.17	0/35	
	hsf-1, let-607 RNAi	8.86 ± 0.38	0/35	<0.001[a]

图序	线虫品系及处理方法	平均寿命 ± SEM/天	非正常死亡数/总数	P值
图3.26（d）重复2	WT, control RNAi	8.65 ± 0.23	0/31	
	WT, let-607 RNAi	14.51 ± 0.33	0/35	<0.001 [a]
	hsf-1, control RNAi	6.91 ± 0.17	0/35	
	hsf-1, let-607 RNAi	9.88 ± 0.44	0/34	<0.001 [a]
图3.26（d）重复3	WT, control RNAi	7.94 ± 0.26	0/35	
	WT, let-607 RNAi	10.74 ± 0.35	0/35	<0.001 [a]
	hsf-1, control RNAi	6.74 ± 0.16	0/35	
	hsf-1, let-607 RNAi	8.06 ± 0.32	0/35	<0.001 [a]
图3.35 重复1	WT, control RNAi	6.17 ± 0.08	0/47	
	WT, let-607 RNAi	7.89 ± 0.09	0/109	<0.001 [a]
	nhr-49, control RNAi	6.08 ± 0.05	0/52	
	nhr-49, let-607 RNAi	7.87 ± 0.15	0/60	<0.001 [a], 0.9612 [b]
图3.35 重复2	WT, control RNAi	6.19 ± 0.07	0/63	
	WT, let-607 RNAi	8.10 ± 0.19	0/61	<0.001 [a]
	nhr-49, control RNAi	6.09 ± 0.05	0/64	
	nhr-49, let-607 RNAi	7.55 ± 0.14	0/53	<0.001 [a], 0.028 [b]
图3.35 重复3	WT, control RNAi	6.08 ± 0.06	0/50	
	WT, let-607 RNAi	7.63 ± 0.20	0/48	<0.001 [a]
	nhr-49, control RNAi	6.07 ± 0.05	0/61	
	nhr-49, let-607 RNAi	7.53 ± 0.13	0/60	<0.001 [a], 0.584 [b]
图3.43 重复1	WT, control RNAi	6.26 ± 0.09	0/54	
	WT, let-607 RNAi	11.69 ± 0.20	13/65	<0.001 [a]
	fat-1, control RNAi	6.26 ± 0.10	0/46	
	fat-1, let-607 RNAi	12.31 ± 0.19	19/58	<0.001 [a], 0.0922 [b]
图3.43 重复2	WT, control RNAi	7.55 ± 0.19	0/44	
	WT, let-607 RNAi	9.21 ± 0.26	0/38	<0.001 [a]
	fat-1, control RNAi	7.54 ± 0.22	0/35	
	fat-1, let-607 RNAi	8.95 ± 0.23	0/40	<0.001 [a], 1 [b]

图序	线虫品系及处理方法	平均寿命 ± SEM/天	非正常死亡数/总数	P值
图3.43 重复3	WT, control RNAi	6.81 ± 0.16	0/37	
	WT, let-607 RNAi	8.42 ± 0.25	0/43	<0.001 [a]
	fat-1, control RNAi	7.05 ± 0.16	0/40	
	fat-1, let-607 RNAi	8.48 ± 0.22	0/50	<0.001 [a], 1 [b]
图3.43 重复1	WT, control RNAi	6.26 ± 0.09	0/54	
	WT, let-607 RNAi	8.78 ± 0.18	0/74	<0.001 [a]
	fat-2, control RNAi	6.11 ± 0.07	0/38	
	fat-2, let-607 RNAi	6.11 ± 0.07	0/37	1 [a], <0.001 [b]
图3.43 重复2	WT, control RNAi	6.41 ± 0.22	0/49	
	WT, let-607 RNAi	8.77 ± 0.27	0/44	<0.001 [a]
	fat-2, control RNAi	6.82 ± 0.24	0/44	
	fat-2, let-607 RNAi	6.96 ± 0.26	0/46	1 [a], <0.001 [b]
图3.43 重复3	WT, control RNAi	7.48 ± 0.19	0/46	
	WT, let-607 RNAi	9.82 ± 0.28	0/55	<0.001 [a]
	fat-2, control RNAi	7.86 ± 0.22	0/57	
	fat-2, let-607 RNAi	8.04 ± 0.22	0/57	1 [a], <0.001 [b]
图3.43 重复1	WT, control RNAi	6.06 ± 0.04	0/63	
	WT, let-607 RNAi	11.52 ± 0.25	0/50	<0.001 [a]
	fat-3, control RNAi	6.17 ± 0.07	0/71	
	fat-3, let-607 RNAi	7.58 ± 0.24	0/104	<0.001 [a], <0.001 [b]
图3.43 重复2	WT, control RNAi	6.65 ± 0.15	0/37	
	WT, let-607 RNAi	9.00 ± 0.25	0/42	<0.001 [a]
	fat-3, control RNAi	6.97 ± 0.16	0/37	
	fat-3, let-607 RNAi	7.91 ± 0.23	0/47	0.0056 [a], 0.0096 [b]
图3.43 重复3	WT, control RNAi	6.83 ± 0.28	0/41	
	WT, let-607 RNAi	11.18 ± 0.21	0/44	<0.001 [a]
	fat-3, control RNAi	7.35 ± 0.25	0/43	
	fat-3, let-607 RNAi	8.26 ± 0.41	0/46	0.0198 [a], <0.001 [b]
图3.43 重复1	WT, control RNAi	6.09 ± 0.06	0/47	
	WT, let-607 RNAi	9.33 ± 0.15	0/63	<0.001 [a]
	fat-4, control RNAi	6.72 ± 0.15	0/39	
	fat-4, let-607 RNAi	9.57 ± 0.12	0/47	<0.001 [a], 1 [b]

图序	线虫品系 及处理方法	平均寿命 ± SEM/天	非正常死 亡数/总数	P 值
图 3.43 重复 2	WT, control RNAi	6.33 ± 0.11	0/43	
	WT, let-607 RNAi	8.51 ± 0.23	0/51	<0.001[a]
	fat-4, control RNAi	6.49 ± 0.14	0/37	
	fat-4, let-607 RNAi	8.98 ± 0.26	0/47	<0.001[a], 0.4566[b]
图 3.43 重复 3	WT, control RNAi	7.10 ± 0.15	0/42	
	WT, let-607 RNAi	9.87 ± 0.30	0/61	<0.001[a]
	fat-4, control RNAi	7.27 ± 0.21	0/33	
	fat-4, let-607 RNAi	9.91 ± 0.26	0/47	<0.001[a], 1[b]
图 3.43 重复 1	WT, control RNAi	6.26 ± 0.09	0/54	
	WT, let-607 RNAi	11.69 ± 0.20	13/65	<0.001[a]
	fat-5, control RNAi	6.00 ± 0.00	0/59	
	fat-5, let-607 RNAi	6.11 ± 0.06	0/55	0.0703[a], <0.001[b]
图 3.43 重复 2	WT, control RNAi	7.55 ± 0.19	0/44	
	WT, let-607 RNAi	9.21 ± 0.26	0/38	<0.001[a]
	fat-5, control RNAi	7.63 ± 0.17	0/54	
	fat-5, let-607 RNAi	7.46 ± 0.18	0/41	1[a], <0.001[b]
图 3.43 重复 3	WT, control RNAi	6.81 ± 0.16	0/37	
	WT, let-607 RNAi	8.42 ± 0.25	0/43	<0.001[a]
	fat-5, control RNAi	7.35 ± 0.15	0/40	
	fat-5, let-607 RNAi	7.42 ± 0.16	0/31	1[a], 0.0074[b]
图 3.47 重复 1	control RNAi	6.03 ± 0.03	0/58	
	let-607 RNAi	11.24 ± 0.26	1/61	<0.001[a]
	asm-3 RNAi	6.72 ± 0.15	0/47	
	asm-3 + let-607 RNAi	9.32 ± 0.19	0/47	<0.001[a], <0.001[b]
图 3.47 重复 2	control RNAi	6.57 ± 0.14	0/56	
	let-607 RNAi	9.51 ± 0.32	0/41	<0.001[a]
	asm-3 RNAi	7.63 ± 0.20	6/43	
	asm-3 + let-607 RNAi	9.62 ± 0.25	3/52	<0.001[a], <0.001[b]
图 3.47 重复 3	control RNAi	7.24 ± 0.17	0/50	
	let-607 RNAi	10.75 ± 0.32	0/59	<0.001[a]
	asm-3 RNAi	7.52 ± 0.18	0/63	
	asm-3 + let-607 RNAi	8.82 ± 0.23	0/44	<0.001[a], <0.001[b]
图 3.47 重复 1	control	6.00 ± 0.00	0/60	
	30 μmol/L desipramine	6.11 ± 0.06	0/54	0.0655[a]

图序	线虫品系及处理方法	平均寿命 ± SEM/天	非正常死亡数/总数	P值
图3.47 重复2	control	6.80 ± 0.17	0/35	
	30 μmol/L desipramine	6.94 ± 0.17	0/36	0.5425[a]
图3.47 重复3	control	8.22 ± 0.25	0/37	
	30 μmol/L desipramine	8.15 ± 0.26	0/41	0.9814[a]
图3.51 重复1	control RNAi	6.03 ± 0.03	0/58	
	let-607 RNAi	11.24 ± 0.26	1/61	<0.001[a]
	sms-5 RNAi	6.72 ± 0.15	0/47	
	sms-5 + let-607 RNAi	9.32 ± 0.19	0/47	<0.001[a], <0.001[b]
图3.51 重复2	control RNAi	6.57 ± 0.14	0/56	
	let-607 RNAi	9.51 ± 0.32	0/41	<0.001[a]
	sms-5 RNAi	7.63 ± 0.20	6/43	
	sms-5 + let-607 RNAi	9.62 ± 0.25	3/52	<0.001[a], <0.001[b]
图3.51 重复3	control RNAi	7.24 ± 0.17	0/50	
	let-607 RNAi	10.75 ± 0.32	0/59	<0.001[a]
	sms-5 RNAi	7.52 ± 0.18	0/63	
	sms-5 + let-607 RNAi	8.82 ± 0.23	0/44	<0.001[a], <0.001[b]
图3.77（a）重复1	WT, control RNAi	6.69 ± 0.18	0/35	
	WT, let-607 RNAi	10.34 ± 0.31	0/35	<0.001[a]
	itr-1, control RNAi	6.82 ± 0.24	0/34	
	itr-1, let-607 RNAi	7.54 ± 0.27	0/35	0.069[a], <0.001[b]
图3.77（a）重复2	WT, control RNAi	6.63 ± 0.16	6/35	
	WT, let-607 RNAi	9.39 ± 0.32	5/35	<0.001[a]
	itr-1, control RNAi	6.69 ± 0.16	4/35	
	itr-1, let-607 RNAi	6.97 ± 0.20	0/35	0.308[a], <0.001[b]
图3.77（a）重复3	WT, control RNAi	7.28 ± 0.16	0/36	
	WT, let-607 RNAi	10.65 ± 0.26	4/35	<0.001[a]
	itr-1, control RNAi	7.37 ± 0.16	0/35	
	itr-1, let-607 RNAi	8.00 ± 0.18	0/35	0.012[a], <0.001[b]
图3.80 重复1	control RNAi	6.65 ± 0.13	0/49	
	let-607 RNAi	10.00 ± 0.20	0/37	<0.001[a]
	egl-8 RNAi	7.27 ± 0.17	0/44	
	egl-8 + let-607 RNAi	7.12 ± 0.19	0/34	0.548[a]

图序	线虫品系及处理方法	平均寿命 ± SEM/天	非正常死亡数/总数	P值
图3.80 重复2	control RNAi	7.20 ± 0.17	0/40	
	let-607 RNAi	9.00 ± 0.20	0/48	<0.001[a]
	egl-8 RNAi	7.71 ± 0.16	6/42	
	egl-8 + let-607 RNAi	6.56 ± 0.13	3/61	0.342[a]
图3.80 重复3	control RNAi	6.14 ± 0.08	0/42	
	let-607 RNAi	8.72 ± 0.26	0/50	<0.001[a]
	egl-8 RNAi	6.57 ± 0.13	0/46	
	egl-8 + let-607 RNAi	6.09 ± 0.06	0/44	0.003[a]
图3.92（a）重复1	WT, control RNAi	6.19 ± 0.07	0/63	
	WT, let-607 RNAi	8.10 ± 0.19	0/61	<0.001[a]
	sgk-1, control RNAi	6.00 ± 0.00	0/53	
	sgk-1, let-607 RNAi	6.08 ± 0.06	0/49	0.1394[a]
图3.92（a）重复2	WT, control RNAi	6.43 ± 0.12	0/47	
	WT, let-607 RNAi	7.80 ± 0.17	0/51	<0.001[a]
	sgk-1, control RNAi	6.20 ± 0.08	0/50	
	sgk-1, let-607 RNAi	6.23 ± 0.08	0/62	0.8270[a]
图3.92（a）重复3	WT, control RNAi	6.17 ± 0.08	0/47	
	WT, let-607 RNAi	7.89 ± 0.09	0/109	<0.001[a]
	sgk-1, control RNAi	6.00 ± 0.00	0/64	
	sgk-1, let-607 RNAi	6.11 ± 0.06	0/55	0.0595[a]
3.92（b）重复1	control RNAi	7.26 ± 0.13	0/54	
	let-607 RNAi	9.03 ± 0.15	0/64	<0.001[a]
	sgk-1 RNAi	6.20 ± 0.09	0/49	
	sgk-1 + let-607 RNAi	6.72 ± 0.14	2/47	0.003[a]
图3.92（b）重复2	control RNAi	6.48 ± 0.14	0/46	
	let-607 RNAi	10.85 ± 0.23	0/40	<0.001[a]
	sgk-1 RNAi	6.19 ± 0.09	6/42	
	sgk-1 + let-607 RNAi	7.22 ± 0.25	3/36	<0.001[a]
图3.92（b）重复3	control RNAi	6.57 ± 0.14	0/42	
	let-607 RNAi	8.95 ± 0.18	0/44	<0.001[a]
	sgk-1 RNAi	6.25 ± 0.10	0/40	
	sgk-1 + let-607 RNAi	7.17 ± 0.24	0/36	<0.001[a]

图序	线虫品系及处理方法	平均寿命 ± SEM/天	非正常死亡数/总数	P值
图3.95 重复1	WT, control RNAi	6.19 ± 0.07	0/63	
	WT, let-607 RNAi	8.10 ± 0.19	0/61	<0.001[a]
	rict-1, control RNAi	6.43 ± 0.13	0/42	
	rict-1, let-607 RNAi	8.42 ± 0.18	0/66	<0.001[a], 0.6903[b]
图3.95 重复2	WT, control RNAi	6.17 ± 0.08	0/47	
	WT, let-607 RNAi	7.89 ± 0.09	0/109	<0.001[a]
	rict-1, control RNAi	8.04 ± 0.14	0/54	
	rict-1, let-607 RNAi	8.18 ± 0.24	0/55	<0.001[a], 0.3679[b]
图3.95 重复3	WT, control RNAi	7.51 ± 0.12	0/53	
	WT, let-607 RNAi	10.32 ± 0.26	0/44	<0.001[a]
	rict-1, control RNAi	9.78 ± 0.24	0/45	
	rict-1, let-607 RNAi	11.59 ± 0.18	0/64	<0.001[a], 0.0005[b]

注：① [a] 表示该组P值为相同处理组内或相同线虫品系间比较；
② [b] 表示该组P值为与野生型线虫let-607 RNAi比较。

F.5 热应激实验统计结果

重复1为文中用图数据,其余为生物学重复。

图序	线虫品系及处理方法	平均寿命 ± SEM/天	非正常死亡数/总数	P值
图3.9(b)重复1	control RNAi	7.37 ± 0.19	0/35	
	let-607 RNAi	10.69 ± 0.28	0/35	<0.001[a]
图3.9(b)重复2	control RNAi	7.26 ± 0.16	0/35	
	let-607 RNAi	9.60 ± 0.24	0/35	<0.001[a]
图3.9(b)重复3	control RNAi	7.00 ± 0.16	0/38	
	let-607 RNAi	9.58 ± 0.16	0/43	<0.001[a]
图3.10重复1	VP303, control RNAi	7.54 ± 0.13	0/39	
	VP303, let-607 RNAi	9.71 ± 0.36	0/34	<0.001[a]
图3.10重复2	VP303, control RNAi	7.58 ± 0.15	2/38	
	VP303, let-607 RNAi	9.08 ± 0.20	0/52	<0.001[a]
图3.10重复3	VP303, control RNAi	8.10 ± 0.15	0/41	
	VP303, let-607 RNAi	10.43 ± 0.30	0/37	<0.001[a]
图3.10重复1	DCL569, control RNAi	7.19 ± 0.17	0/32	
	DCL569, let-607 RNAi	7.26 ± 0.16	0/38	0.748[a]
图3.10重复2	DCL569, control RNAi	9.00 ± 0.21	0/40	
	DCL569, let-607 RNAi	9.38 ± 0.18	0/39	0.302[a]
图3.10重复3	DCL569, control RNAi	7.85 ± 0.08	0/41	
	DCL569, let-607 RNAi	8.00 ± 0.19	0/37	0.393[a]
图3.10重复1	WM118, control RNAi	5.94 ± 0.06	0/34	
	WM118, let-607 RNAi	5.95 ± 0.05	0/37	0.952[a]
图3.10重复2	WM118, control RNAi	6.97 ± 0.17	0/35	
	WM118, let-607 RNAi	7.14 ± 0.17	0/35	0.476[a]
图3.10重复3	WM118, control RNAi	6.97 ± 0.16	0/37	
	WM118, let-607 RNAi	6.86 ± 0.15	0/44	0.625[a]
图3.10重复1	NR222, control RNAi	8.32 ± 0.23	0/37	
	NR222, let-607 RNAi	8.48 ± 0.27	0/33	0.652[a]

附 录

图序	线虫品系及处理方法	平均寿命 ± SEM/天	非正常死亡数/总数	P值
图3.10 重复2	NR222, control RNAi	9.18 ± 0.29	0/39	
	NR222, let-607 RNAi	9.88 ± 0.31	0/33	0.085[a]
图3.10 重复3	NR222, control RNAi	8.67 ± 0.25	0/33	
	NR222, let-607 RNAi	8.75 ± 0.21	0/32	0.906[a]
图3.13（b）重复1	control RNAi	8.08 ± 0.20	3/59	
	let-607 RNAi	10.64 ± 0.31	7/54	<0.001[a]
图3.13（b）重复2	control RNAi	7.61 ± 0.21	7/51	
	let-607 RNAi	9.46 ± 0.27	8/56	<0.001[a]
图3.13（b）重复3	control RNAi	7.20 ± 0.18	13/45	
	let-607 RNAi	8.77 ± 0.28	15/65	<0.001[a]
图3.13（b）重复4	control RNAi	8.49 ± 0.22	7/51	
	let-607 RNAi	10.13 ± 0.33	1/52	<0.001[a]
图3.26（b）重复1	WT, control RNAi	8.60 ± 0.29	0/30	
	WT, let-607 RNAi	10.71 ± 0.34	0/34	<0.001[a]
	daf-16, control RNAi	7.70 ± 0.17	0/33	
	daf-16, let-607 RNAi	9.06 ± 0.25	0/32	<0.001[a], <0.001[b]
图3.26（b）重复2	WT, control RNAi	9.35 ± 0.18	0/34	
	WT, let-607 RNAi	13.94 ± 0.41	2/34	<0.001[a]
	daf-16, control RNAi	7.78 ± 0.10	0/36	
	daf-16, let-607 RNAi	10.88 ± 0.49	0/32	<0.001[a], <0.001[b]
图3.26（b）重复3	WT, control RNAi	8.76 ± 0.22	0/34	
	WT, let-607 RNAi	11.12 ± 0.30	0/34	<0.001[a]
	daf-16, control RNAi	7.20 ± 0.18	0/35	
	daf-16, let-607 RNAi	8.34 ± 0.29	0/35	<0.001[a], <0.001[b]
图3.26（c）重复1	WT, control RNAi	8.76 ± 0.22	0/34	
	WT, let-607 RNAi	11.12 ± 0.30	0/34	<0.001[a]
	hsf-1, control RNAi	8.29 ± 0.22	0/34	
	hsf-1, let-607 RNAi	8.88 ± 0.28	0/34	0.060[a], <0.001[b]

图序	线虫品系及处理方法	平均寿命 ± SEM/天	非正常死亡数/总数	P值
图3.26（c）重复2	WT, control RNAi	9.60 ± 0.14	0/35	
	WT, let-607 RNAi	13.16 ± 0.23	0/38	<0.001[a]
	hsf-1, control RNAi	9.20 ± 0.20	0/35	
	hsf-1, let-607 RNAi	10.88 ± 0.32	0/43	<0.001[a], <0.001[b]
图3.26（c）重复3	WT, control RNAi	9.53 ± 0.19	0/34	
	WT, let-607 RNAi	14.97 ± 0.27	0/35	<0.001[a]
	hsf-1, control RNAi	9.37 ± 0.28	0/38	
	hsf-1, let-607 RNAi	11.26 ± 0.50	0/38	<0.001[a], <0.001[b]
图3.43 重复1	WT, control RNAi	8.03 ± 0.08	0/59	
	WT, let-607 RNAi	9.71 ± 0.16	0/42	<0.001[a]
	fat-1, control RNAi	7.62 ± 0.13	0/52	
	fat-1, let-607 RNAi	9.28 ± 0.16	0/53	<0.001[a], 0.0768[b]
图3.43 重复2	WT, control RNAi	7.38 ± 0.16	0/39	
	WT, let-607 RNAi	9.56 ± 0.27	0/36	<0.001[a]
	fat-1, control RNAi	8.05 ± 0.18	0/40	
	fat-1, let-607 RNAi	9.03 ± 0.25	0/37	0.0005[a], 0.1507[b]
图3.43 重复3	WT, control RNAi	6.76 ± 0.16	0/37	
	WT, let-607 RNAi	8.00 ± 0.19	0/39	<0.001[a]
	fat-1, control RNAi	7.25 ± 0.17	0/32	
	fat-1, let-607 RNAi	8.22 ± 0.21	0/37	0.001[a], 0.418[b]
图3.43 重复1	WT, control RNAi	7.93 ± 0.09	0/60	
	WT, let-607 RNAi	11.01 ± 0.37	0/43	<0.001[a]
	fat-2, control RNAi	8.58 ± 0.21	0/64	
	fat-2, let-607 RNAi	7.53 ± 0.32	0/43	0.0338[a], <0.001[b]
图3.43 重复2	WT, control RNAi	6.86 ± 0.25	0/35	
	WT, let-607 RNAi	9.91 ± 0.34	0/43	<0.001[a]
	fat-2, control RNAi	8.30 ± 0.30	0/40	
	fat-2, let-607 RNAi	7.68 ± 0.34	0/38	0.2475[a], <0.001[b]

图序	线虫品系及处理方法	平均寿命 ± SEM/天	非正常死亡数/总数	P值
图3.43（c）重复3	WT, control RNAi	7.48 ± 0.19	0/46	
	WT, let-607 RNAi	9.82 ± 0.28	0/55	<0.001[a]
	fat-2, control RNAi	7.86 ± 0.22	0/57	
	fat-2, let-607 RNAi	8.04 ± 0.22	0/57	0.2144[a], 0.0024[b]
图3.43 重复1	WT, control RNAi	7.90 ± 0.05	0/62	
	WT, let-607 RNAi	10.25 ± 0.25	0/72	<0.001[a]
	fat-3, control RNAi	7.99 ± 0.21	0/47	
	fat-3, let-607 RNAi	9.35 ± 0.36	0/55	0.0011[a], 0.1091[b]
图3.43 重复2	WT, control RNAi	7.38 ± 0.16	0/39	
	WT, let-607 RNAi	9.56 ± 0.27	0/36	<0.001[a]
	fat-3, control RNAi	8.00 ± 0.23	0/32	
	fat-3, let-607 RNAi	9.57 ± 0.27	0/47	0.0001[a], 0.7953[b]
图3.43 重复3	WT, control RNAi	7.53 ± 0.24	0/34	
	WT, let-607 RNAi	9.76 ± 0.30	0/42	<0.001[a]
	fat-3, control RNAi	8.75 ± 0.28	0/40	
	fat-3, let-607 RNAi	10.57 ± 0.29	0/49	<0.001[a], 0.0546[b]
图3.43 重复1	WT, control RNAi	8.47 ± 0.14	0/60	
	WT, let-607 RNAi	12.69 ± 0.25	0/61	<0.001[a]
	fat-4, control RNAi	9.19 ± 0.16	0/37	
	fat-4, let-607 RNAi	10.86 ± 0.28	0/49	<0.001[a], <0.001[b]
图3.43 重复2	WT, control RNAi	7.30 ± 0.16	0/37	
	WT, let-607 RNAi	9.96 ± 0.22	0/49	<0.001[a]
	fat-4, control RNAi	7.55 ± 0.18	0/31	
	fat-4, let-607 RNAi	8.85 ± 0.24	0/40	<0.001[a], <0.001[b]
图3.43 重复3	WT, control RNAi	8.76 ± 0.22	0/42	
	WT, let-607 RNAi	10.92 ± 0.27	0/50	<0.001[a]
	fat-4, control RNAi	8.18 ± 0.26	0/33	
	fat-4, let-607 RNAi	9.49 ± 0.25	0/43	<0.001[a], <0.001[b]

图序	线虫品系及处理方法	平均寿命 ± SEM/天	非正常死亡数/总数	P值
图3.43 重复1	WT, control RNAi	8.03 ± 0.08	0/59	
	WT, let-607 RNAi	9.71 ± 0.16	0/42	<0.001[a]
	fat-5, control RNAi	8.19 ± 0.08	0/53	
	fat-5, let-607 RNAi	10.00 ± 0.15	0/63	0.1856[a], <0.001[b]
图3.43 重复2	WT, control RNAi	7.38 ± 0.19	0/39	
	WT, let-607 RNAi	9.56 ± 0.26	0/36	<0.001[a]
	fat-5, control RNAi	7.56 ± 0.17	0/32	
	fat-5, let-607 RNAi	9.27 ± 0.18	0/52	0.6222[a], <0.001[b]
图3.43 重复3	WT, control RNAi	6.76 ± 0.16	0/37	
	WT, let-607 RNAi	8.00 ± 0.19	0/39	<0.001[a]
	fat-5, control RNAi	6.67 ± 0.17	0/39	
	fat-5, let-607 RNAi	7.94 ± 0.12	0/36	0.7061[a], <0.001[b]
图3.47 重复1	control RNAi	7.56 ± 0.13	0/41	
	let-607 RNAi	10.16 ± 0.39	0/25	<0.001[a]
	asm-3 RNAi	7.03 ± 0.20	0/37	
	asm-3 + let-607 RNAi	9.52 ± 0.36	0/42	<0.001[a], 1[b]
图3.47 重复2	control RNAi	8.51 ± 0.21	0/39	
	let-607 RNAi	11.19 ± 0.26	0/42	<0.001[a]
	asm-3 RNAi	9.12 ± 0.21	0/34	
	asm-3 + let-607 RNAi	10.44 ± 0.34	0/36	<0.001[a], 0.1603[b]
图3.47 重复3	control RNAi	7.32 ± 0.16	0/47	
	let-607 RNAi	9.86 ± 0.31	0/58	<0.001[a]
	asm-3 RNAi	7.38 ± 0.16	0/39	
	asm-3 + let-607 RNAi	9.07 ± 0.28	0/56	<0.001[a], 0.1374[b]
图3.47 重复1	control	7.89 ± 0.06	0/56	
	30 μmol/L desipramine	7.62 ± 0.11	0/53	0.0304[a]
图3.47 重复2	control	8.59 ± 0.25	0/32	
	30 μmol/L desipramine	8.82 ± 0.23	0/39	0.5698[a]
图3.47 重复3	control	7.44 ± 0.15	0/36	
	30 μmol/L desipramine	7.14 ± 0.18	0/37	0.1977[a]

注：① [a] 表示该组P值为相同处理组内或相同线虫品系间比较；
② [b] 表示该组P值为与野生型线虫let-607 RNAi比较。

F.6 内质网应激实验统计结果

重复1为文中用图数据,其余为生物学重复。

图序	线虫品系及处理方法	平均寿命 ± SEM/天	非正常死亡数/总数	P值
图3.9(c) 重复1	control RNAi	6.21 ± 0.20	3/56	
	let-607 RNAi	8.89 ± 0.32	6/51	<0.001[a]
图3.9(c) 重复2	control RNAi	6.62 ± 0.22	3/65	
	let-607 RNAi	8.90 ± 0.29	5/56	<0.001[a]
图3.9(c) 重复3	control RNAi	7.00 ± 0.18	6/59	
	let-607 RNAi	9.43 ± 0.30	6/55	<0.001[a]
图3.9(c) 重复4	control RNAi	6.07 ± 0.18	1/72	
	let-607 RNAi	8.62 ± 0.26	11/70	<0.001[a]

注:[a]表示该组P值为相同处理组内或相同线虫品系间比较。

F.7 寿命实验统计结果

重复1为文中用图数据，其余为生物学重复。

图序	线虫品系 及处理方法	平均寿命 ± SEM/天	非正常死 亡数/总数	P值
图3.9（d） 重复1	WT, control RNAi	16.39 ± 0.62	2/55	
	WT, *let*-607 RNAi	24.95 ± 0.97	0/55	<0.001[a]
图3.9（d） 重复2	WT, control RNAi	16.23 ± 0.62	1/55	
	WT, *let*-607 RNAi	25.09 ± 0.87	0/55	<0.001[a]
图3.9（d） 重复3	WT, control RNAi	15.89 ± 0.67	0/55	
	WT, *let*-607 RNAi	26.52 ± 0.87	3/55	<0.001[a]
图3.9（d） 重复4	WT, control RNAi	16.33 ± 0.65	0/55	
	WT, *let*-607 RNAi	25.38 ± 0.91	0/55	<0.001[a]
图3.13（c） 重复1	WT, control RNAi	16.63 ± 0.65	5/59	
	WT, *let*-607 RNAi	22.10 ± 0.78	6/64	<0.001[a]
图3.13（c） 重复2	WT, control RNAi	17.76 ± 0.59	5/63	
	WT, *let*-607 RNAi	20.66 ± 0.75	0/61	<0.001[a]
图3.13（c） 重复3	WT, control RNAi	18.42 ± 0.52	2/64	
	WT, *let*-607 RNAi	21.79 ± 0.74	3/60	<0.001[a]
图3.13（c） 重复4	WT, control RNAi	17.50 ± 0.58	3/63	
	WT, *let*-607 RNAi	21.75 ± 0.84	6/62	<0.001[a]
图3.27 重复1	WT, control RNAi	18.69 ± 0.99	11/58	
	WT, *let*-607 RNAi	24.79 ± 1.08	2/55	<0.001[a]
	daf-16, control RNAi	13.85 ± 0.40	2/55	
	daf-16, *let*-607 RNAi	15.05 ± 0.54	4/49	0.022[a], <0.001[b]
图3.27 重复2	WT, control RNAi	17.93 ± 0.85	10/60	
	WT, *let*-607 RNAi	25.93 ± 1.09	1/53	<0.001[a]
	daf-16, control RNAi	13.29 ± 0.51	4/55	
	daf-16, *let*-607 RNAi	14.80 ± 0.67	4/49	0.008[a], <0.001[b]
图3.27 重复3	WT, control RNAi	17.93 ± 0.60	6/61	
	WT, *let*-607 RNAi	22.76 ± 0.96	4/60	<0.001[a]
	daf-16, control RNAi	13.59 ± 0.33	6/62	
	daf-16, *let*-607 RNAi	15.41 ± 0.41	4/62	<0.001[a], <0.001[b]

附 录

图序	线虫品系及处理方法	平均寿命 ± SEM/天	非正常死亡数/总数	P 值
图 3.27 重复 4	WT, control RNAi	19.10 ± 0.62	8/57	
	WT, let-607 RNAi	23.19 ± 0.86	5/59	<0.001 [a]
	daf-16, control RNAi	13.86 ± 0.36	11/62	
	daf-16, let-607 RNAi	15.17 ± 0.39	4/62	0.016 [a], <0.001 [b]
图 3.28 (a) 重复 1	WT, control RNAi	18.87 ± 0.57	0/60	
	WT, let-607 RNAi	25.90 ± 0.97	0/60	<0.001 [a]
	daf-2, control RNAi	33.63 ± 1.49	0/59	
	daf-2, let-607 RNAi	39.10 ± 1.65	0/60	0.014 [a]
图 3.28 (a) 重复 2	WT, control RNAi	18.93 ± 0.58	0/60	
	WT, let-607 RNAi	26.69 ± 0.77	0/61	<0.001 [a]
	daf-2, control RNAi	33.00 ± 1.34	0/60	
	daf-2, let-607 RNAi	40.27 ± 1.65	0/60	<0.001 [a]
图 3.28 (a) 重复 3	WT, control RNAi	18.21 ± 0.70	0/57	
	WT, let-607 RNAi	24.54 ± 1.00	0/59	<0.001 [a]
	daf-2, control RNAi	31.70 ± 1.55	0/60	
	daf-2, let-607 RNAi	36.98 ± 1.68	0/61	0.006 [a]
图 3.28 (a) 重复 4	WT, control RNAi	18.04 ± 0.61	0/56	
	WT, let-607 RNAi	24.60 ± 1.02	0/57	<0.001 [a]
	daf-2, control RNAi	31.56 ± 1.47	0/59	
	daf-2, let-607 RNAi	37.69 ± 1.76	0/59	0.001 [a]
图 3.28 (b) 重复 1	WT, control RNAi	17.47 ± 0.63	3/50	
	WT, let-607 RNAi	27.68 ± 0.92	4/54	<0.001 [a]
	glp-1, control RNAi	34.67 ± 1.20	3/56	
	glp-1, let-607 RNAi	34.27 ± 1.22	3/56	0.862 [a]
图 3.28 (b) 重复 2	WT, control RNAi	17.23 ± 0.67	3/54	
	WT, let-607 RNAi	25.44 ± 0.99	3/55	<0.001 [a]
	glp-1, control RNAi	33.01 ± 1.27	2/54	
	glp-1, let-607 RNAi	34.60 ± 1.25	3/55	0.376 [a]

图序	线虫品系及处理方法	平均寿命 ± SEM/天	非正常死亡数/总数	P值
图3.28 (b) 重复3	WT, control RNAi	17.55 ± 0.51	2/51	
	WT, let-607 RNAi	23.48 ± 0.98	3/52	<0.001[a]
	glp-1, control RNAi	30.20 ± 1.52	5/55	
	glp-1, let-607 RNAi	34.01 ± 1.33	6/55	0.142[a]
图3.28 (b) 重复4	WT, control RNAi	17.99 ± 0.65	8/55	
	WT, let-607 RNAi	23.85 ± 1.02	6/57	<0.001[a]
	glp-1, control RNAi	30.62 ± 1.44	6/55	
	glp-1, let-607 RNAi	31.59 ± 1.37	6/54	0.906[a]
图3.29 (a) 重复1	WT, control RNAi	15.44 ± 0.54	2/66	
	WT, let-607 RNAi	25.16 ± 1.01	5/60	<0.001[a]
	daf-9, control RNAi	13.30 ± 0.47	0/60	
	daf-9, let-607 RNAi	19.67 ± 0.63	2/63	<0.001[a]
图3.29 (a) 重复2	WT, control RNAi	15.38 ± 0.51	0/68	
	WT, let-607 RNAi	24.07 ± 0.87	6/62	<0.001[a]
	daf-9, control RNAi	12.50 ± 0.39	1/61	
	daf-9, let-607 RNAi	18.78 ± 0.62	3/62	<0.001[a]
图3.29 (a) 重复3	WT, control RNAi	14.82 ± 0.50	1/69	
	WT, let-607 RNAi	24.69 ± 0.92	7/62	<0.001[a]
	daf-9, control RNAi	12.27 ± 0.40	1/60	
	daf-9, let-607 RNAi	18.07 ± 0.72	1/60	<0.001[a]
图3.29 (a) 重复4	WT, control RNAi	15.08 ± 0.53	0/63	
	WT, let-607 RNAi	23.28 ± 1.09	2/60	<0.001[a]
	daf-9, control RNAi	13.05 ± 0.39	0/63	
	daf-9, let-607 RNAi	19.07 ± 0.61	2/60	<0.001[a]
图3.29 (b) 重复1	WT, control RNAi	17.38 ± 0.49	1/62	
	WT, let-607 RNAi	25.97 ± 0.95	5/65	<0.001[a]
	daf-12, control RNAi	13.81 ± 0.45	0/62	
	daf-12, let-607 RNAi	19.82 ± 0.71	5/61	<0.001[a]

图序	线虫品系及处理方法	平均寿命 ± SEM/天	非正常死亡数/总数	P值
图3.29 (b) 重复2	WT, control RNAi	17.43 ± 0.49	0/63	
	WT, let-607 RNAi	24.34 ± 0.87	3/63	<0.001[a]
	daf-12, control RNAi	13.82 ± 0.47	0/66	
	daf-12, let-607 RNAi	20.00 ± 0.64	3/66	<0.001[a]
图3.29 (b) 重复3	WT, control RNAi	16.83 ± 0.48	0/63	
	WT, let-607 RNAi	25.15 ± 0.84	4/65	<0.001[a]
	daf-12, control RNAi	14.44 ± 0.45	0/64	
	daf-12, let-607 RNAi	20.76 ± 0.80	3/61	<0.001[a]
图3.29 (b) 重复4	WT, control RNAi	16.32 ± 0.52	2/65	
	WT, let-607 RNAi	23.92 ± 0.94	3/63	<0.001[a]
	daf-12, control RNAi	14.10 ± 0.48	0/63	
	daf-12, let-607 RNAi	18.54 ± 0.77	1/60	<0.001[a]
图3.29 (c) 重复1	WT, control RNAi	17.70 ± 0.46	1/60	
	WT, let-607 RNAi	24.11 ± 0.87	7/62	<0.001[a]
	tcer-1, control RNAi	20.64 ± 0.72	1/60	
	tcer-1, let-607 RNAi	27.36 ± 0.98	3/62	<0.001[a]
图3.29 (c) 重复2	WT, control RNAi	16.81 ± 0.51	1/63	
	WT, let-607 RNAi	24.81 ± 0.92	4/61	<0.001[a]
	tcer-1, control RNAi	19.94 ± 0.75	0/62	
	tcer-1, let-607 RNAi	26.67 ± 1.19	4/61	<0.001[a]
图3.29 (c) 重复3	WT, control RNAi	16.97 ± 0.49	2/66	
	WT, let-607 RNAi	26.00 ± 0.92	3/62	<0.001[a]
	tcer-1, control RNAi	19.10 ± 0.70	0/60	
	tcer-1, let-607 RNAi	29.84 ± 1.01	1/63	<0.001[a]
图3.29 (c) 重复4	WT, control RNAi	17.02 ± 0.51	0/63	
	WT, let-607 RNAi	23.07 ± 0.88	3/65	<0.001[a]
	tcer-1, control RNAi	21.15 ± 0.70	0/66	
	tcer-1, let-607 RNAi	27.21 ± 1.12	3/61	<0.001[a]

图序	线虫品系及处理方法	平均寿命 ± SEM/天	非正常死亡数/总数	P值
图3.29 (d) 重复1	WT, control RNAi	16.55 ± 0.44	5/60	
	WT, let-607 RNAi	23.43 ± 0.85	4/60	<0.001[a]
	kri-1, control RNAi	15.37 ± 0.55	3/66	
	kri-1, let-607 RNAi	22.77 ± 0.89	1/61	<0.001[a]
图3.29 (d) 重复2	WT, control RNAi	14.95 ± 0.49	2/61	
	WT, let-607 RNAi	22.44 ± 0.90	2/61	<0.001[a]
	kri-1, control RNAi	13.27 ± 0.57	5/59	
	kri-1, let-607 RNAi	21.76 ± 1.00	2/61	<0.001[a]
图3.29 (d) 重复3	WT, control RNAi	14.90 ± 0.48	3/65	
	WT, let-607 RNAi	20.23 ± 1.00	3/64	<0.001[a]
	kri-1, control RNAi	13.37 ± 0.54	2/62	
	kri-1, let-607 RNAi	21.23 ± 1.09	1/61	<0.001[a]
图3.29 (d) 重复4	WT, control RNAi	12.69 ± 0.51	0/61	
	WT, let-607 RNAi	22.33 ± 1.00	2/62	<0.001[a]
	kri-1, control RNAi	14.20 ± 0.53	0/61	
	kri-1, let-607 RNAi	21.23 ± 1.03	1/63	<0.001[a]
图3.52 重复1	control RNAi	15.89 ± 0.67	0/55	
	let-607 RNAi	26.52 ± 0.87	3/55	<0.001[a]
	sms-5 RNAi	17.38 ± 0.84	0/55	
	sms-5 + let-607 RNAi	23.26 ± 0.97	0/54	<0.001[a], 0.174[b]
图3.52 重复2	control RNAi	16.23 ± 0.62	1/55	
	let-607 RNAi	25.09 ± 0.87	0/55	<0.001[a]
	sms-5 RNAi	18.69 ± 0.84	0/55	
	sms-5 + let-607 RNAi	23.28 ± 0.88	0/58	<0.001[a], 0.181[b]
图3.52 重复3	control RNAi	16.39 ± 0.62	2/55	
	let-607 RNAi	24.95 ± 0.97	0/55	<0.001[a]
	sms-5 RNAi	18.37 ± 0.89	1/55	
	sms-5 + let-607 RNAi	23.12 ± 0.94	0/57	<0.001[a], 0.034[b]

图序	线虫品系及处理方法	平均寿命 ± SEM/天	非正常死亡数/总数	P值
图3.52 重复4	control RNAi	16.33 ± 0.65	0/55	
	let-607 RNAi	26.19 ± 0.83	3/55	<0.001[a]
	sms-5 RNAi	17.27 ± 0.94	0/55	
	sms-5 + let-607 RNAi	23.60 ± 0.99	0/55	<0.001[a], 0.116[b]
图3.77 (b) 重复1	WT, control RNAi	16.39 ± 0.62	2/55	
	WT, let-607 RNAi	24.95 ± 0.97	0/55	<0.001[a]
	itr-1, control RNAi	13.79 ± 0.37	2/55	
	itr-1, let-607 RNAi	17.16 ± 0.66	0/55	<0.001[a], <0.001[b]
图3.77 (b) 重复2	WT, control RNAi	16.23 ± 0.62	1/55	
	WT, let-607 RNAi	25.09 ± 0.87	0/55	<0.001[a]
	itr-1, control RNAi	13.71 ± 0.41	0/55	
	itr-1, let-607 RNAi	17.38 ± 0.65	0/55	<0.001[a], <0.001[b]
图3.77 (b) 重复3	WT, control RNAi	15.89 ± 0.67	0/55	
	WT, let-607 RNAi	26.52 ± 0.87	3/55	<0.001[a]
	itr-1, control RNAi	13.85 ± 0.36	0/55	
	itr-1, let-607 RNAi	17.93 ± 0.78	0/60	<0.001[a], <0.001[b]
图3.77 (b) 重复4	WT, control RNAi	16.33 ± 0.65	0/55	
	WT, let-607 RNAi	25.38 ± 0.91	0/55	<0.001[a]
	itr-1, control RNAi	14.15 ± 0.38	0/55	
	itr-1, let-607 RNAi	17.60 ± 0.75	0/56	<0.001[a], <0.001[b]
图3.93 重复1	control RNAi	16.39 ± 0.62	2/55	
	let-607 RNAi	24.95 ± 0.97	0/55	<0.001[a]
	sgk-1 RNAi	14.92 ± 0.44	2/55	
	sgk-1 + let-607 RNAi	15.16 ± 0.39	0/55	0.838[a]
图3.93 重复2	control RNAi	16.23 ± 0.62	1/55	
	let-607 RNAi	25.09 ± 0.87	0/55	<0.001[a]
	sgk-1 RNAi	15.13 ± 0.38	0/55	
	sgk-1 + let-607 RNAi	14.91 ± 0.39	0/55	0.790[a]

图序	线虫品系 及处理方法	平均寿命 ± SEM/天	非正常死 亡数/总数	P 值
图 3.93 重复 3	control RNAi	15.89 ± 0.67	0/55	
	let-607 RNAi	26.52 ± 0.87	3/55	<0.001 [a]
	sgk-1 RNAi	14.18 ± 0.47	0/55	
	sgk-1 + let-607 RNAi	14.62 ± 0.41	0/60	0.781 [a]
图 3.93 重复 4	control RNAi	16.33 ± 0.65	0/55	
	let-607 RNAi	25.38 ± 0.91	0/55	<0.001 [a]
	sgk-1 RNAi	14.25 ± 0.49	0/55	
	sgk-1 + let-607 RNAi	13.96 ± 0.55	0/57	0.925 [a]

注：① [a] 表示该组 P 值为相同处理组内或相同线虫品系间比较；
② [b] 表示该组 P 值为与野生型线虫 let-607 RNAi 比较。